226
Publications
52

RESTRICTED

The information given in this document is not to be communicated, either directly or indirectly to the Press or to any person not authorized to receive it.

HANDBOOK OF ENEMY AMMUNITION

PAMPHLET No. 13

GERMAN ROCKETS, GUN AND MORTAR AMMUNITION

By Command of the Army Council,

THE WAR OFFICE,
  17*th October*, 1944.

LONDON:

HANDBOOK OF ENEMY AMMUNITION

CONTENTS TABLE

German Ammunition

	Page
Push igniter DZ 35 (Druckzünder 35)	2
Fuze F.P.Z8001	3
Fuze A.Z. 5095	4
Fuze Hbgr. Z. 35 K.	5
Fuze le Jgr. Z.23 nA. for use under Ballistic Cap.	8
Fuze Electric for ignition of rockets (E.R.Z.39)	11
Fuze Wgr. Z. 50 +	13
Fuze Kz. 38 with gaine	16
Fuze Bd. Z. DOV.	18
Gaine Model 40 B (Zündladung 40B)	20
Gaine Model 41 (Zündladung 41)	22
Gaine 36 Np. (Zündladung 36 Np.)	25
Hollow charge bomb with projector. Faustpatrone 2. (Panzerfaust 30)	25
8 cm. Mortar H.E. bombs 38 umg and 39 umg.	30
7·3 Propaganda rocket	31
8·8 cm. Anti-tank hollow charge rocket. (R. Pz. B. Gr. 4322)	34
15 cm. H.E. Rocket (15 cm. Wurfgranate 41 Spreng)	40
21 cm. H.E.B.C. Rocket (21 cm. Wurfgranate 42 Spr.)	43
30 cm. H.E. Rocket (30 cm. Wurfkörper Spreng)	49
4 cm. Cartridge Q.F. H.E. (Bofors type) (4 cm. Sprgr Patr. 28 Flak)	54
8·8 cm. Flak 41 Cartridge Q.F. A.P.C.B.C./T. (8·8 cm. Pz gr. Patr. 39 Flak 41)	60
12·8 cm. Flak 40 Cartridge Q.F. H.E. fuzed Zt Z.S/30 (12·8 cm. Sprgr. Patr. L/4·5)	62

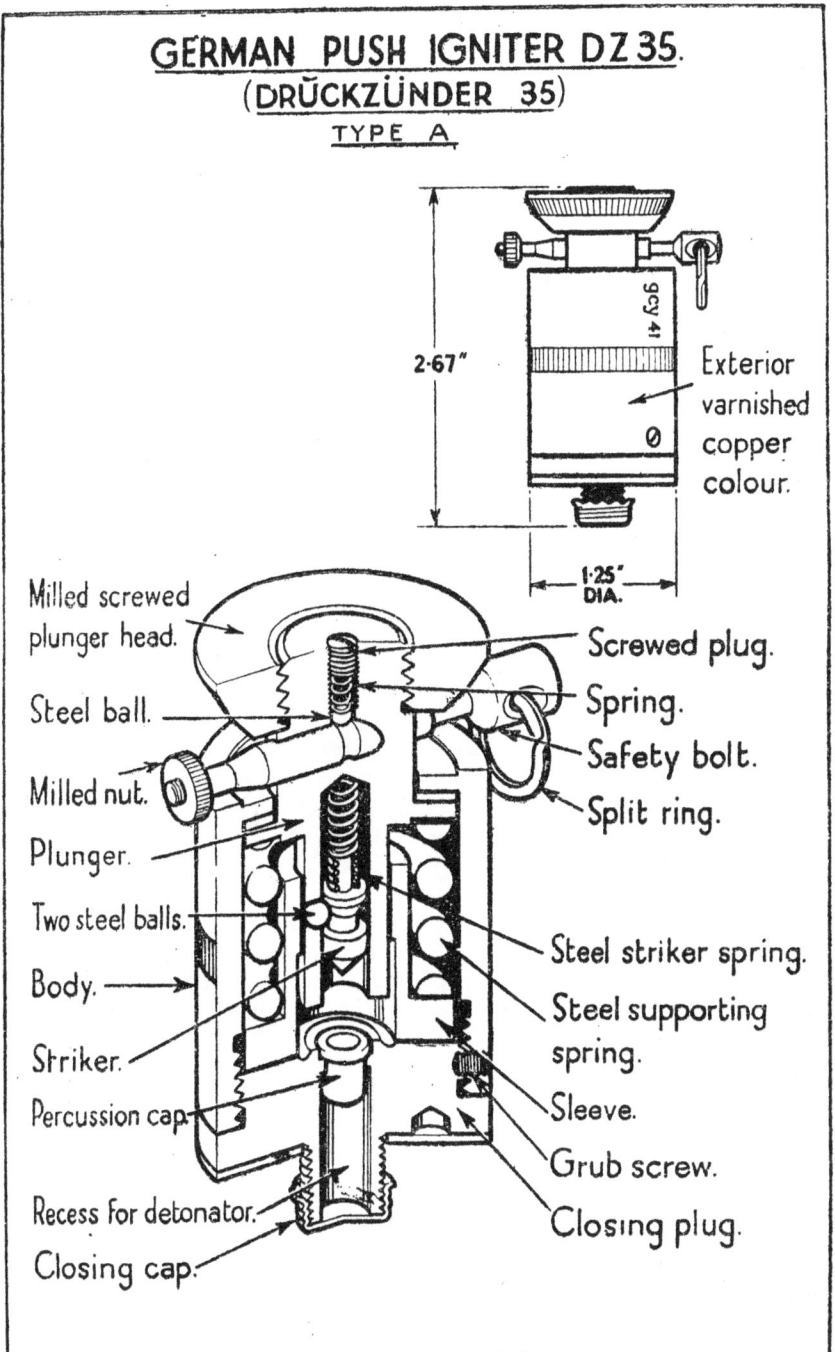

Fig. 1

HANDBOOK OF ENEMY AMMUNITION
PAMPHLET No. 13

GERMAN PUSH IGNITER D.Z.35
(Druckzünder 35)
(Fig. 1)

The D.Z.35 push igniter is cylindrical in shape, and designed for use with improvised mines, etc. The safety arrangement is a safety bolt which passes diametrically through the external portion of the plunger, thereby preventing the latter being pushed in to fire a cap. It is reported that a pressure of between 130 lb. and 165 lb. is required to operate the igniter. A specimen tested, functioned at a pressure of approximately 150 lb. The overall length is 2·67 inches and the diameter of the body 1·25 inches. It is varnished a copper colour.

Another type of D.Z.35, made of brass, similar in action but rather different in construction and with a body diameter of 1 inch, is reported to be in existence. The pressure necessary to fire this type is considerably less than that required to function the one described above.

The aluminium body is a cylindrical casing shaped at the top to form an internal flange, and screwthreaded internally at its lower end to receive an aluminium closing plug with central channel containing the percussion cap. The central channel extends into a projection formed in the base of the plug for the insertion of a detonator when required. The projection is screwthreaded externally for insertion in the charge container and is closed by a screwed transit cap. The plug, when screwed home, is secured by a grub screw which passes through the wall of the cylindrical body. A thin cardboard washer fits on the underside of the plug and surrounds the projection.

Above the plug, is a sleeve with an external flange at its base end which forms a seating for the lower end of the steel spiral supporting spring surrounding its smaller diameter. Internally, the sleeve at its upper end acts as a guide to the plunger, its lower end is of greater diameter to allow the balls to be forced outwards when pressure is applied to the plunger.

The upper end of the supporting spring bears against the external flange of an aluminium plunger which is assembled from the inside of the casing, so that its top end protrudes from the case, whilst the lower end fits into the sleeve. The upper side of the flange bears against the underside of the internal flange at the top of the case and limits its upward movement. The protruding portion of the plunger is screwthreaded externally to receive a flat circular screwed head with milled edge. An aluminium safety bolt with split ring at one end and a retaining nut at the other, passes through a horizontal channel in the plunger below the milled head. A groove, semi-circular in section, is formed in the centre of the bolt to engage a spring-loaded ball located in a vertical channel in the upper end of the plunger. The ball and spring are retained in position by a screwed plug.

The plunger is recessed from the base to accommodate the striker and its steel spring, and has two radial holes each locating a steel ball which engage a circumferential groove in the striker.

The steel striker has a stem formed at its upper end, over which its steel spiral spring is assembled. The spring is held under compression between a flange on the striker and the upper end of the recess. A circumferential groove, with the upper side inclined, is formed in the striker body to engage with the two steel balls.

Action

To arm the igniter the safety bolt is withdrawn by removing its retaining nut and pulling the bolt out of the plunger against the resistance of the spring-loaded ball.

When pressure is applied to the plunger it is forced into the body against the resistance of the supporting spring. When forced down sufficiently, the steel balls lose the support of the smaller diameter in the sleeve and are pushed outwards by the inclined surface of the striker groove under pressure from the striker spring, which also drives the striker on the cap assembly.

GERMAN FUZE (F.P. Z8001) for H.C. ANTI-TANK BOMB

(Faustpatrone or Panzerfaust)

(Fig. 2)

The main parts are accommodated in a cylindrical steel container, closed by a steel washer and consist of an inertia pellet with needle, arming collar, three springs, detonator, and detonator holder. All, except the detonator, are made of steel.

The inertia pellet and arming collar are hollow cylinders with an external flange at one end to form bearings for a steel spring surrounding them. Both have free longitudinal movement inside the container, the arming collar being arranged to move telescopically over the inertia pellet. The other end of the inertia pellet is closed except for three flash holes and has the needle and a distance piece secured to it.

The distance piece is made in two diameters. It supports a spiral spring, whilst the fuze is in the unarmed condition, and a wire locking spring, hexagonal in shape, is fitted around the smaller diameter.

The detonator holder and closing washer are supported on a step at the end of the container and secured by turning over the continuation of it.

Action

On acceleration, the arming collar sets back so that the spiral and hexagonal springs expand. The hexagonal spring locks the arming collar to the inertia pellet and the whole is free to move forward through the expanded spiral spring when the flight of the projectile is checked.

German Fuze F.P.Z. 8001 for H.C. Anti-tank Bomb.
(Faustpatrone.)

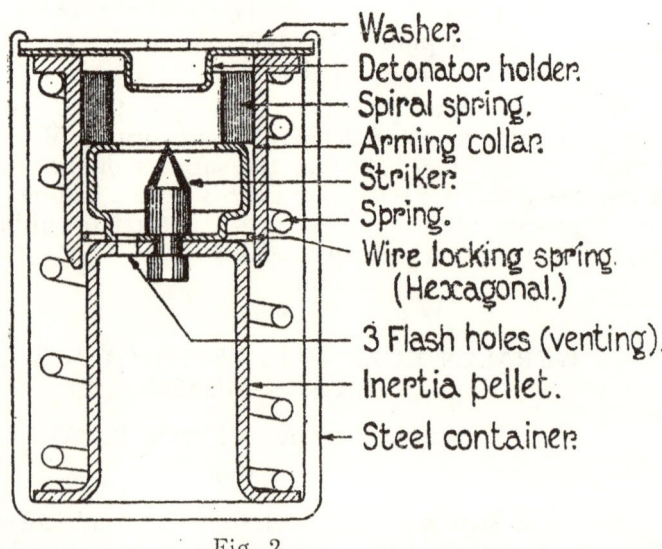

Fig. 2

GERMAN FUZE A.Z.5095

(Fig. 3)

This fuze is used in the nose of the German 8·8 cm. A.Tk. Hollow Charge Rocket Projectile. Except for the undermentioned differences, it is identical with the Fuze 5075 described in Pamphlet No. 11, page 26.

(i) The body is made of steel.
(ii) The striker is made of black plastic.
(iii) The retaining spring supporting the arming sleeve has only two arms, instead of four, and these are thinner to permit the fuze to be armed at a much smaller acceleration.
(iv) The spring holder is cut in two places to accommodate the arms.
(v) The magazine holder, which is of steel, screws into the fuze body direct. The adapter has been omitted.
(vi) A safety pin is provided. This passes through the fuze body and a circumferential groove cut in the arming collar.

The weight of the fuze is 3·02 ozs. The diameter of the threads is 1·057 inches and the overall length 2·12 inches.

GERMAN FUZE Hbgr Z 35 K

(Fig. 4)

This fuze, which may be identified by the stamping Hbgr. Z. 35 K on the body, is of the direct action and graze type with an optional delay of 0·10 secs. It is designed for use under a ballistic cap and is used in the 21 cm. H.E. rocket and other projectiles with ballistic caps.

The fuze mechanism is substantially the same in design and dimensions as the AZ 23 umg 0,15 described in Pamphlet No. 12, except that a wooden hammer and closing disc are removed and the following arrangement substituted, to enable the fuze to be used under a ballistic cap.

The recess in the top of the setting head is screwthreaded to receive an adapter which screws on to a metal disc and is secured by a setscrew. The disc closes the bottom of the recess and supports the base end of a rod forming an extended hammer.

The adapter is bored centrally and has a coned opening at the top.

The hammer is a tube of light alloy with a solid extension piece, slightly less in diameter, attached at each end. The lower end of the hammer is housed in the fuze adapter whilst the upper end is held by another adapter in the nose of the ballistic cap and retained by a brass closing disc.

Action

The fuze is normally set at non-delay for transport, and may be set for delay by the aid of a special fuze key.

The action of the fuze is the same as that described in Pamphlet No. 12 for the AZ 23 umg 0,15, except that at direct action the ballistic cap is shattered and the hammer forces the needle on to the detonator.

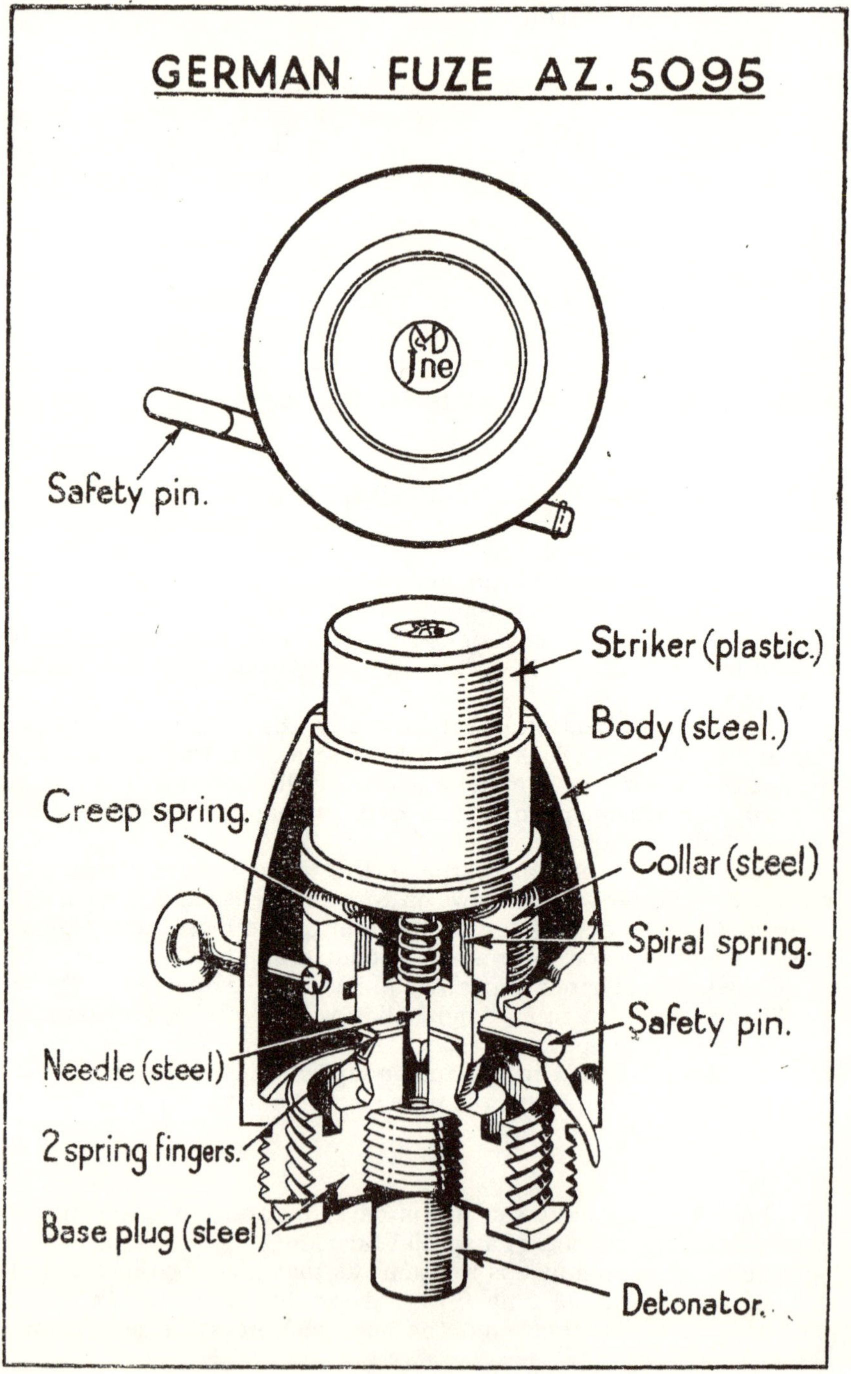

GERMAN FUZE AZ.5095

Safety pin.

Striker (plastic.)
Body (steel.)
Creep spring.
Collar (steel)
Spiral spring.
Safety pin.
Needle (steel)
2 spring fingers.
Base plug (steel)
Detonator.

Fig. 3

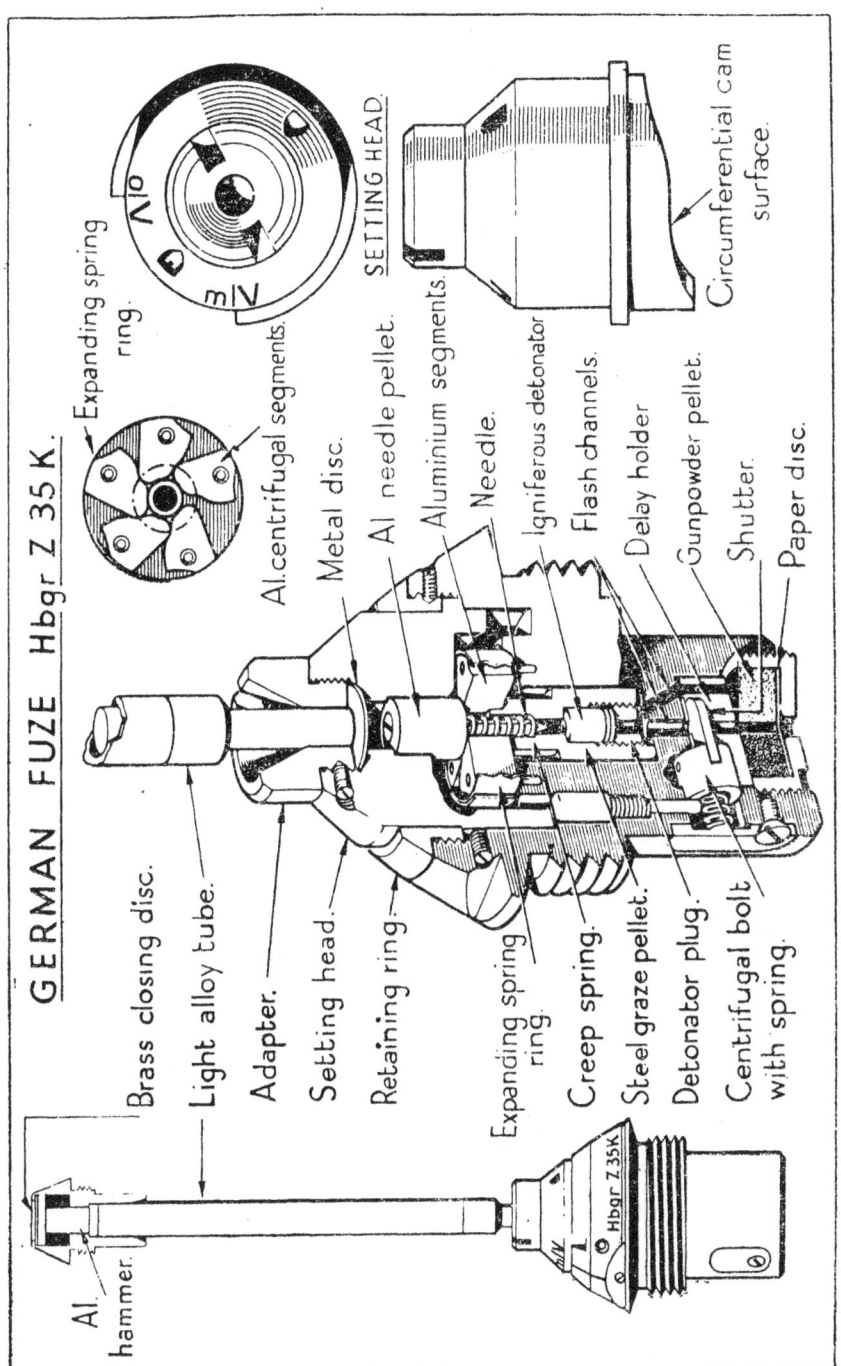

Fig. 4

GERMAN PERCUSSION FUZE le Jgr Z 23 nA
for use under ballistic caps

(Fig. 5)

This fuze, which consists of a slightly modified le Jgr Z 23 nA with the addition of a wooden striker extension rod and nose bush to fit into the ballistic cap, is used in projectiles with ballistic caps. It is being replaced by the Hbgr Z 35 K, also described in this pamphlet.

It is designed to function on impact through a wooden extension rod or on graze and has an optional delay of 0·15 seconds.

The fuze consists of an aluminium body, needle, needle holder, creep spring, centrifugal segments, expanding spring ring, detonator pellet, detonator collar, delay mechanism and a magazine.

The body is in three sections, to facilitate the assembly of the mechanism which are screwed together and secured by spinning.

The nose section is bored centrally in two diameters to form a chamber at the base to accommodate the needle holder, and above it, a recess to accommodate a wooden rod forming an extension to the striker. The top of the recess is coned outwards. Externally at the bottom it is screwthreaded for insertion in the centre section. The steel cylindrical needle holder is bored centrally in two diameters to receive the steel needle. The needle is secured in the holder by turning the metal at the top of the boring over the head of the needle. The underside of the holder is recessed to form a bearing for one end of the creep spring which surrounds the stem of the needle.

The centre section is bored centrally in two diameters, screw-threaded internally at the top to receive the nose section and externally at the bottom for insertion in the base section. The chamber formed in the top houses five brass centrifugal segments each pivoting on a pin. The segments are surrounded by an expanding spring ring which, at rest, prevents the needle and detonator pellet moving towards each other to fire the detonator. The lower chamber houses the detonator pellet.

The detonator pellet is of steel, cylindrical in shape, and smaller in external diameter at its forward end ; it is bored centrally in three diameters to form an internal flange separating two chambers. The upper chamber receives the base end of the creep spring and the point of the needle, and the lower chamber the detonator. The detonator is retained in the pellet by a light alloy plug with central fire channel. The plug is secured by a steel pin. The base of the plug is coned and located in an iron inertia collar. Paper washers are inserted between the plug and detonator.

The base section of the fuze body is bored, centrally to form chambers for the collar, delay holder and magazine, and radially to accommodate the delay mechanism. It is screwthreaded inter-

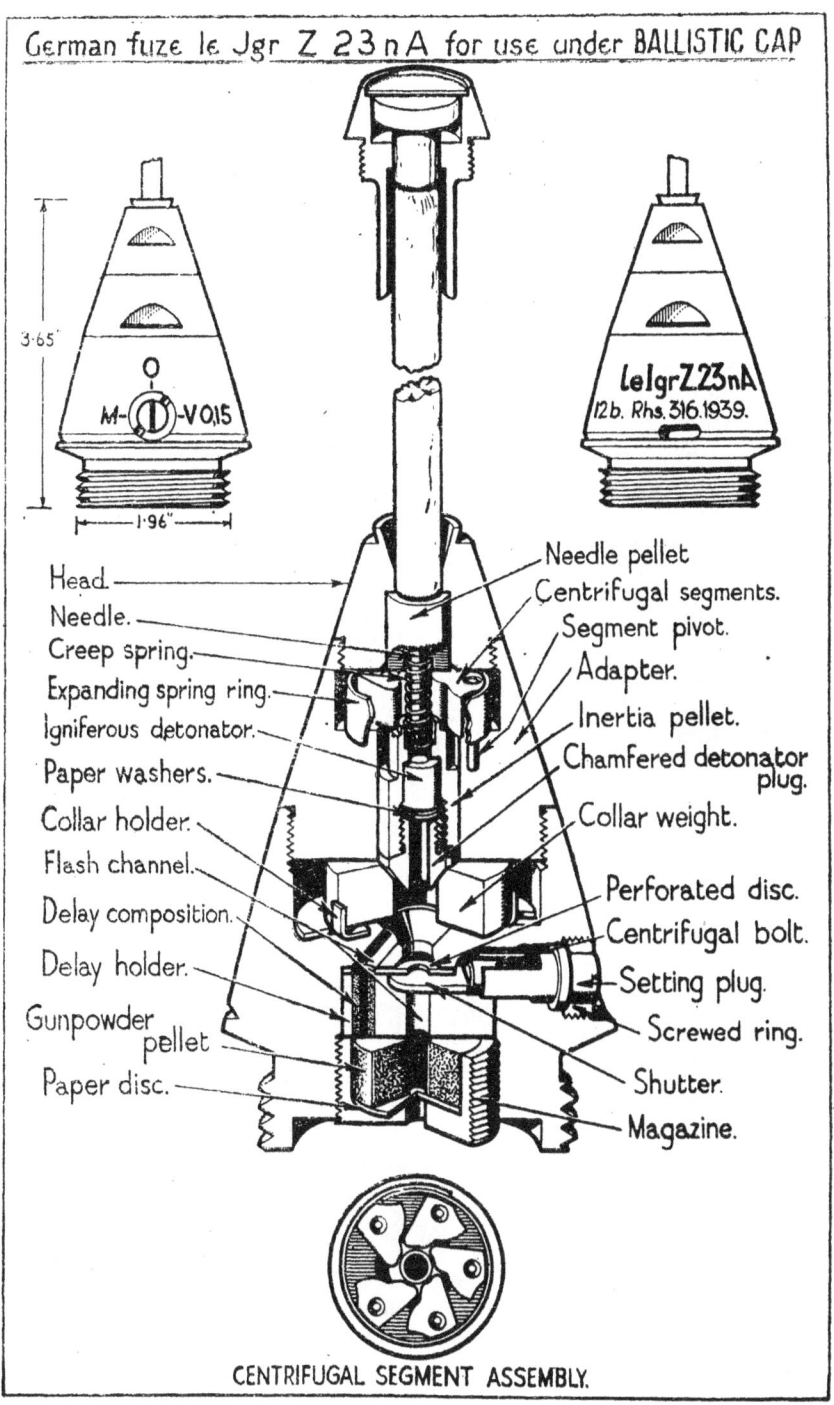

Fig. 5

nally at the top to receive the centre section and externally at the base below the flange for insertion in the shell. The chamber at the top houses an iron inertia ring rustproofed by some process which leaves a matt surface. The ring is square in section and coned internally at the top to match the coned portion of the plug in the detonator pellet. The diameter of the ring is considerably less than that of the chamber which houses it. The ring is centred by five upturned lugs of a rustproofed steel washer which fits closely at the bottom of the chamber. The chamber at the bottom houses the delay holder above a magazine which closes the base of the fuze.

The delay holder is an aluminium pellet with a central flash channel, and another channel, displaced from the centre, filled with delay composition. Both channels are in communication with the magazine filling and there is an inclined channel between the delay composition and the centre of the fuze. A recess in the top surface of the holder accommodates a centrifugal shutter. The delay composition, in a similar fuze, consisted of sulphur 9·2 per cent., potassium nitrate 72·6 per cent., charcoal 18·2 per cent.

The magazine is screwthreaded externally and contains a perforated pressed pellet of gunpowder. A flash hole in the base of the magazine is closed by a paper disc inserted below the pellet.

The central flash channel in the delay holder may be closed or opened by an optional delay mechanism. The delay mechanism consists of a setting plug, centrifugal bolt, spiral spring, plate and screwed ring. The setting plug is tubular steel closed at one end and has a slot on the outside which serves as a setting index. The plug is retained in the fuze body by the steel screwed ring. The cavity in the plug houses a cylindrical steel centrifugal bolt and spiral spring. When the fuze is set to non-delay, a slot, cut diametrically across the mouth of the cavity, receives a centrifugal copper shutter that slides in a recess in the top surface of the delay holder and unmasks the central fire channel. A brass plate with holes bored to correspond with the delay and central fire is placed on the holder and forms an upper surface for the copper plate.

Action

Before firing

The needle is separated from the detonator by the centrifugal segments, which are retained in the closed position by the expanding spring. The shutter of the delay mechanism closes the central fire channel by pressure from the centrifugal bolt. This position is maintained whether the fuze is set delay on non-delay. The delay channel is always uncovered.

To set the fuze for non-delay action, the slot in the setting plug is turned to a position in prolongation with the axis of the fuze to the marking " O " on the fuze body, thereby bringing the slotted recess in the plug opposite the shutter. For delay action the setting plug is turned at right angles to the axis of the fuze to the marking " M " and " V " on the fuze body, and in this position the plug prevents movement of the shutter during flight, so closing the central fire channel.

During flight—Non-delay action

The centrifugal segments swing outwards one after another, and the surrounding spring expands. During deceleration only the creep spring prevents the detonator pellet carrying the detonator on to the needle. The centrifugal bolt of the delay mechanism moves outwards compressing its spring, and the plate under centrifugal action slides into the plug and opens the central flash channel.

Delay action

When set to delay, the plug prevents the shutter sliding under centrifugal action and the central fire channel remains closed.

On impact

By direct action the extension rod forces the needle directly on to the detonator. On graze, the detonator pellet is carried forward on to the needle, or it may be forced on to the needle by a sideways movement of the inertia collar. The flash from the detonator passes either through the delay channel or through the central channel to the magazine, according to the setting of the fuze, and thence to the gaine in the shell.

GERMAN FUZE ELECTRIC FOR IGNITION OF ROCKETS
(E.R.Z.39)
(Fig. 6)

This electric fuze is provided for use with the 15 cm. and 21 cm. rockets. The overall dimensions and general shape of the fuze are shown in Fig. 6.

The fuze consists essentially of a body, holder, fuze head, two leads, two contacts, priming composition and gunpowder pellets.

The body is of black plastic material, the lower half is cylindrical and the upper half coned and formed with a spigot. It is bored centrally in three diameters to form two chanbers and, in the base, a recess. The top chamber forms a magazine containing three perforated pellets of gunpowder; it is closed by turning the top of the spigot over an aluminium closing disc. The lower chamber accommodates a pellet of wax substance in a tubular cardboard container which forms a holder for the fuze head and the end of two insulated leads. The fuze head is surrounded by a blob of priming composition, composed of a mixture of lead picrate, nitrocellulose, charcoal and potassium nitrate. The recess in the base

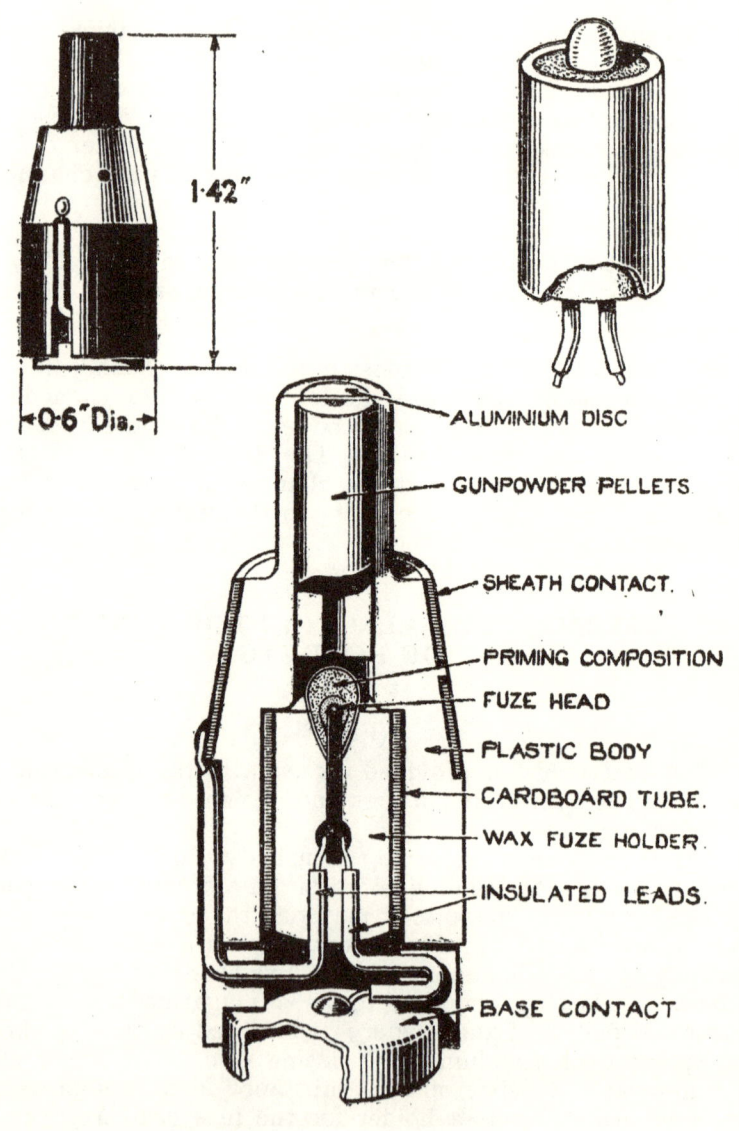

Fig. 6

is closed by an inverted tin plate cup which forms a contact, to which the end of one of the leads is soldered. The other lead is passed through a radial hole and a channel cut in the periphery of the body, the end being soldered to a thin aluminium or tin plate sheath covering the coned surface of the body, which forms the earth return contact.

Action

The electrical circuit is through the base contact, insulated lead, fuze head, return lead and the earth return sheath.

The electrically-heated bridge wire in the fuze head ignites the priming composition, which in turn fires the gunpowder pellets in the magazine. The flash from the magazine ignites the igniter in the rocket.

GERMAN FUZE Wgr Z 50 +
(Fig. 7)

The German fuze Wgr Z 50 + is a direct action and graze fuze with provision for allways action should flight be abnormal, and is used in the 32 cm. incendiary rocket and the 28 cm. and 30 cm. H.E. rockets. It is identified by the marking "Wgr. Z 50 +" stamped in the body above the flange. *The overall length of the fuze with cap is 2 inches and its diameter across the threads is 1·4 inches. It is provided with a cap and safety pin.

The fuze consists of a body, needle, needle holder two centrifugal bolts, expanding spring ring, creep spring, detonator and holder, inertia collar, centreing washer, base plug, cap and safety pin.

The body is of aluminium 2·5 inches in length. Externally, below the flange, it is screwthreaded for insertion in the rocket. Above the flange, it is coned and has a flat top. A circumferential groove is cut in the coned portion to accommodate an expanding spring ring. Internally the body is bored in three diameters, to form chambers 0·6 inch and 1·1 inches in diameter respectively, and an internal flange at the base.

The needle is of steel and provided with a head; it is held in the needle holder by a screwed plug. The needle holder is of aluminium formed with an external flange at its base; it is recessed and screwthreaded to receive the head of the needle and securing plug. The needle and holder are assembled from the inside of the fuze body so that the head of the holder protrudes from the top; its upward movement is limited by the flange on the holder which bears against the internal flange in the body. The needle is retained in this position by a safety pin and two centrifugal bolts and is protected by a cap.

The cap is of thin metal, cone shaped with a flat top; it is pierced horizontally to receive the safety pin.

* Two other types bearing the same marking, are known to be in use. One differs slightly internally; the other has a paper safety ring instead of the safety cap. All function in the same manner.

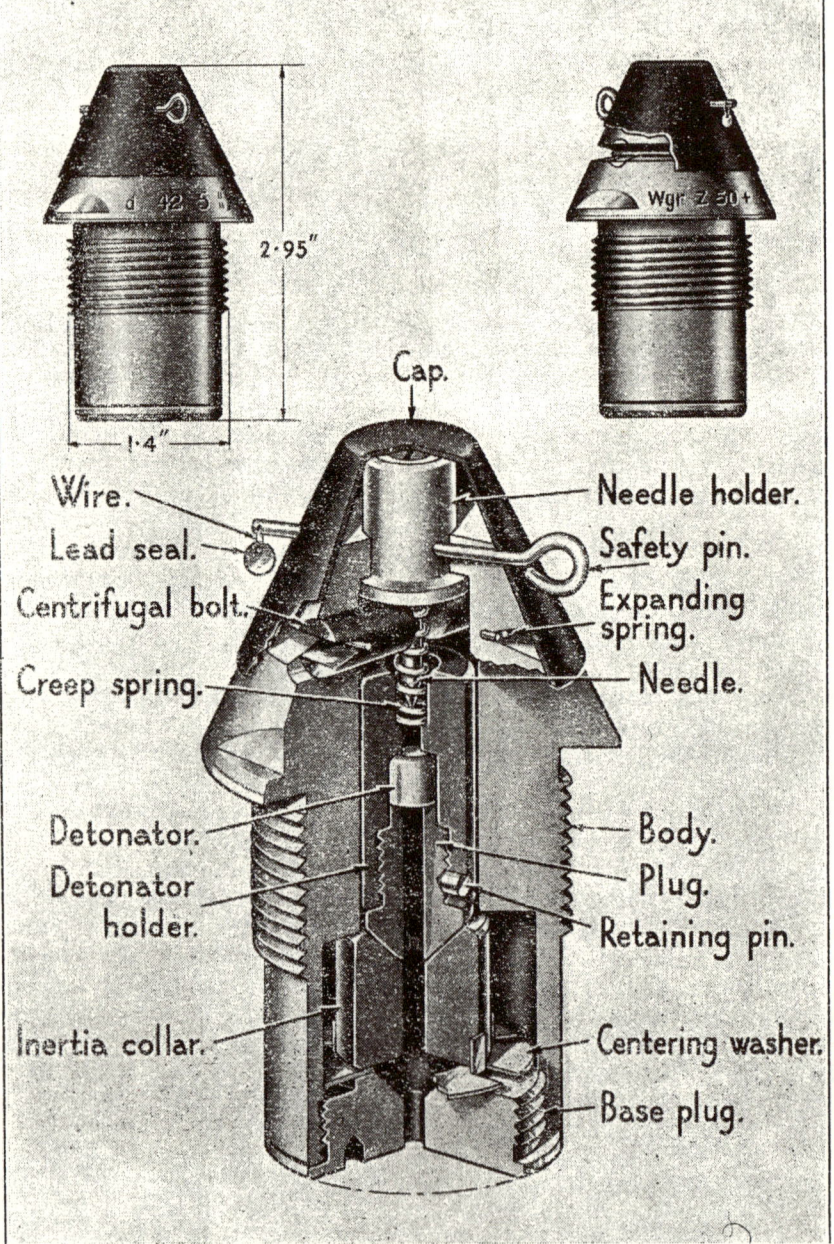

Fig. 7

The safety pin passes through the cap, fuze body and the needle holder. The pin is prevented from falling out by a piece of wire passing through a hole at one end and secured by a lead seal.

The two centrifugal bolts are cylindrical in shape and slotted at one end. They are partially housed in two radial channels and protrude to support the needle holder on its underside, thereby preventing the needle being forced on to the detonator. The bolts are retained in position by a thin expanding spring ring which engages the slots and fits into the circumferential groove in the fuze body.

The detonator holder is of steel, cylindrical in shape and bored in three diameters to form an internal flange separating two recesses. The recess at the top accommodates the point of the needle, and the flange provides a seating for one end of the creep spring which surrounds the needle. The recess below the flange accommodates the detonator, which is secured by a screwed plug with central flash hole. The plug is secured by a retaining pin let into a radial hole in the body of the detonator holder. The base of the plug is coned and fits into an inertia collar.

The inertia collar is of steel, cylindrical in shape and bored centrally to form a flash hole. The flash hole has a coned opening at the top to match and accommodate the coned portion of the plug in the detonator holder. The diameter of the collar is considerably less than that of the chamber which accommodates it. The collar is centred by five upturned strips cut in a brass centring washer which is a close fit in the chamber. The chamber is closed by a screwed base plug with central flash hole.

Action

Before firing

The safety pin is withdrawn and the cap removed. The needle is separated from the detonator by the centrifugal bolts which are retained in the closed position by the expanding spring ring.

After firing

The centrifugal bolts swing outwards under centrifugal action, overcoming the expanding spring ring, thereby permitting the needle and detonator holder to move towards one another. The creep spring prevents creep action. On impact the needle is forced on to the detonator by direct action. On graze the detonator is carried forward on to the needle, or it may be forced on to the needle by a sideways movement of the inertia collar. The flash from the detonator passes through the central fire channel in the inertia collar and base plug to a gaine in the shell, which in turn detonates the shell filling.

GERMAN FUZE Kz 38 WITH GAINE

(Fig. 8)

This fuze, which may be identified by the stencilling "Kz 38" on the dome-shaped body is, including the gaine, 1 oz. in weight and 2 inches in length, the length of the body protruding from the nose of the shell is 0·6 inch. The empty fuze is similar to the British fuze No. 250, Mk. II. The fuze is of the direct action type, and the principal parts are, the body with adapter, striker guide, needle and hammer, arming sleeve and arming spring, stirrup spring and ferrule, three balls, igniferous detonator and gaine.

The dome-shaped body has a flat nose and is prepared with an internal channel stepped in four diameters and screwthreaded internally at its base to receive an adapter. The smaller diameter at the forward end houses the head of a wooden hammer, and below it is a chamber to accommodate the arming sleeve after firing, and a second step which forms a shoulder and bearing surface for the upper end of the ferrule. The channel is closed at its front end by a copper disc 0·006 inch thick.

The adapter is screwthreaded externally to receive the body and to screw into the shell. Internally it is screwthreaded to receive the striker guide and its upper end has a chamber to receive an arming sleeve and arming spring which surround the upper end of the striker guide. The sleeve and spring are retained in the unarmed position by a stirrup spring and ferrule. A pin through the side of the adapter prevents the ferrule from turning.

The striker guide is screwthreaded externally to suit the adapter and to receive a gaine body, and has a central channel. The upper part of the channel forms a guide for the hammer and needle, and the lower part accommodates an igniferous detonator held between a shoulder and a plug, with central fire channel, which screws into the bottom of the channel. Three radial holes are bored in the upper part of the channel and partially accommodate three balls which rest against the underside of the needle head and thereby hold the needle off the detonator. Before firing, they are retained in position by the arming sleeve.

The needle is of steel and flanged at its upper end to enable it to be engaged by the three balls and held in the safe position. A wooden hammer, shaped with a head, rests on the top of the striker and is retained in position by a copper disc, which closes the nose of the fuze.

The arming sleeve is cylindrical and fits around the striker guide to retain the balls. It is flanged at its upper end to form a seating for the arming spring and has a curved periphery to suit the stirrup spring.

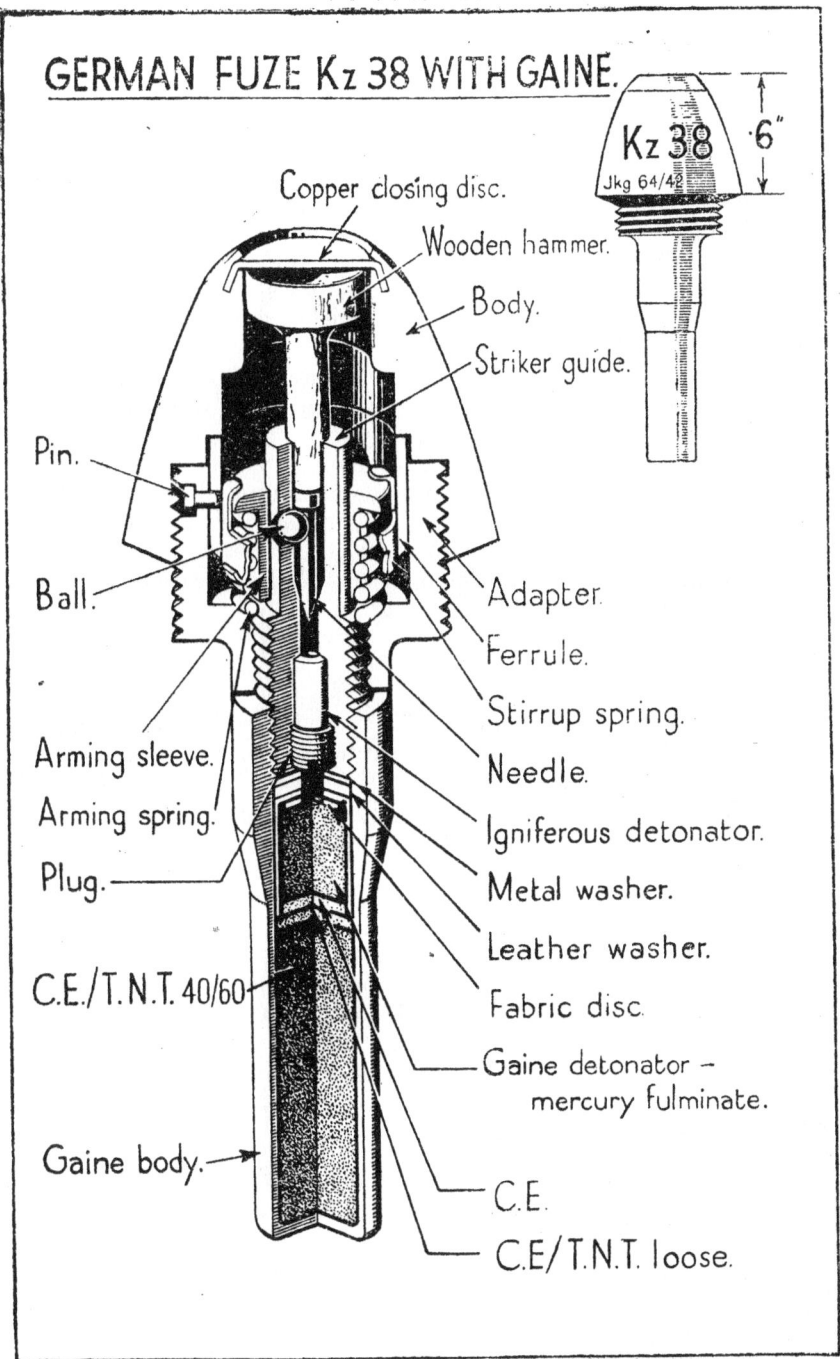

Fig. 8

The stirrup spring is cylindrical, its lower edge has a number of external projections which fit under the ferrule, and its upper edge has a number of internal projections which fit over the arming sleeve and keep the arming spring under compression.

The ferrule is cylindrical, its upper end engages a shoulder in the body and its lower end the stirrup spring. A pin engages a slot cut vertically in one side of the ferrule and prevents its rotary movement in the fuze.

The igniferous detonator contains about 0·5 grains of a mixture of mercury fulminate, potassium chlorate, antimony sulphide (with, possibly, a small proportion of ground glass), followed by a thin layer of gunpowder. It is the same size as the ignitory detonator in Fuze No. 250.

The gaine, approximately 1·05 inches long, is a steel cylindrical body closed at its bottom end and containing 7 grains of CE/TNT (40/60) pressed extremely hard into the base with a thin layer of the same composition in the form of loose crystals on top. Above this is an inverted cup-shaped capsule containing a detonator composition consisting of 0·46 grains of C.E. under 6 grains of fulminate of mercury. The flash hole in the top of the capsule is closed by a fabric disc on its underside. The mouth of the gaine is screwthreaded to enable it to be screwed on the striker guide of the fuze, and contains a leather washer under a metal washer which fits over the detonator cap.

Action

This fuze is of the floating needle type and is not provided with a supporting spring; it, therefore, cannot be over-emphasized that before being fired the closing disc in the nose of the fuze should be examined to see that it is not damaged or perforated, otherwise the round may be fired prematurely by air pressure acting directly on the hammer.

On firing, the ferrule sets back and takes the stirrup spring with it, thereby releasing the arming sleeve, which is forced upwards by its spring and unmasks the three holes in the striker guide.

During flight, the needle and hammer tend to creep forward owing to deceleration, the three balls are released and fly outwards under centrifugal action and so release the needle.

On impact, the hammer is forced in driving the needle into the detonator, which is fired and in turn sets off the gaine detonator and gaine which detonates the shell filling.

GERMAN BASE FUZE Bd. Z. DOV.
FROM 15 cm. H.E. ROCKET
(Fig. 9)

This is a direct action base fuze used in the 15 cm. H.E. rocket. It may be identified by the stamping " Bd. Z. DOV " in the base. The overall length of the fuze is 1·4 inches and its maximum diameter, over the threads, is 1·75 inches.

GERMAN BASE FUZE Bdz DOV.

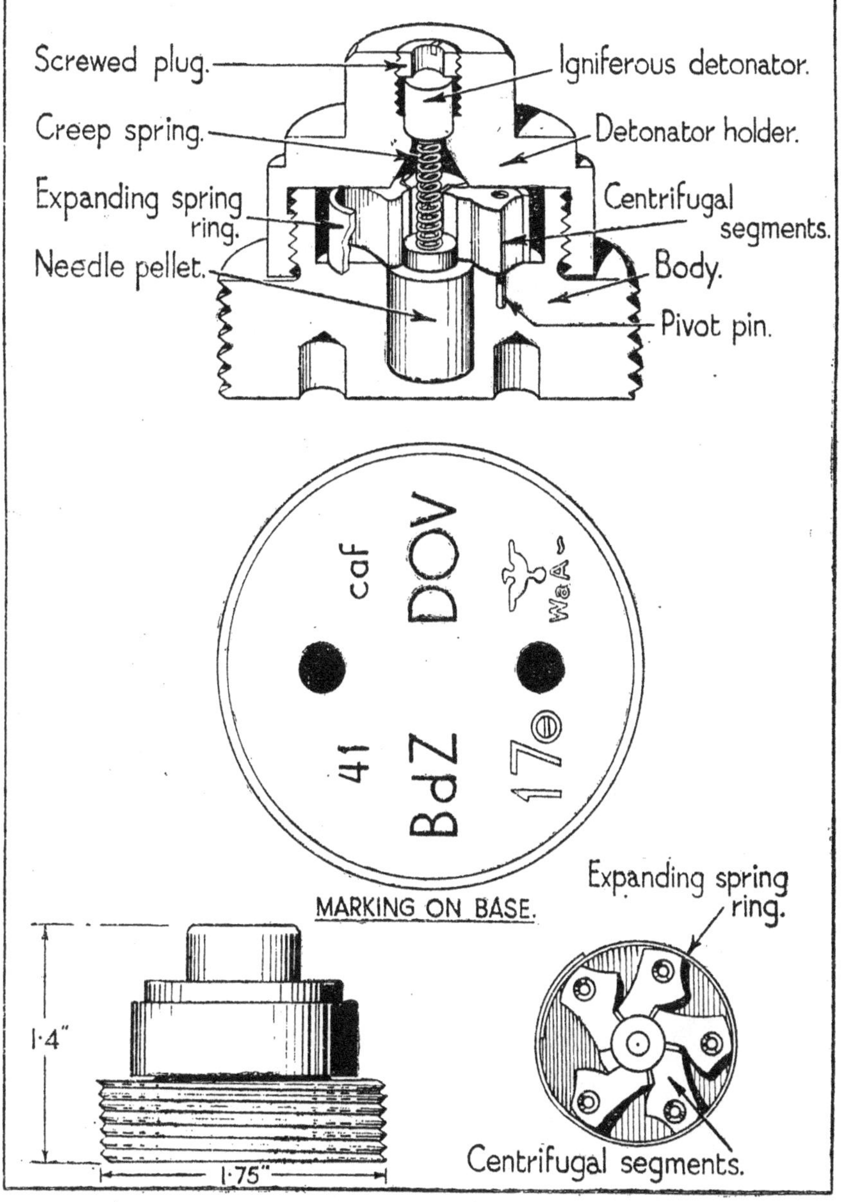

- Screwed plug.
- Creep spring.
- Expanding spring ring.
- Needle pellet.
- Igniferous detonator.
- Detonator holder.
- Centrifugal segments.
- Body.
- Pivot pin.

MARKING ON BASE.

caf DOV W&A
4f BdZ 17

Expanding spring ring.
Centrifugal segments.

1·4"
1·75"

Fig. 9

The fuze consist of a steel body, steel needle, five centrifugal segments of light alloy, expanding spring ring, creep spring, steel detonator holder, igniferous detonator and steel screwed plug. All steel components are rustproofed.

The body is screwthreaded externally with a left-hand thread for insertion in the base of the rocket, and the forward end is reduced in diameter and screwthreaded for the attachment of the detonator holder. It is bored centrally from the front in two diameters to form chambers to house the needle pellet and centrifugal segments.

The needle is integral with the inertia pellet. The pellet is housed in the base of the body, and provided with two shoulders below the needle, which is located centrally in the larger chamber.

The five centrifugal segments are pivoted on pins and are shaped so that one locks the other. In the unarmed position, the segments engage the larger shoulder of the needle pellet and hold the needle off the detonator. They are surrounded and retained in this position by an expanding spring ring.

The spiral creep spring surrounding the needle is held between the smaller shoulder of the needle pellet and a seating in the detonator holder.

The detonator holder is bored centrally to form two chambers, which are connected by a coned boring with a seating for one end of the creep spring. The larger chamber is screwthreaded to receive the forward end of the fuze body. The smaller chamber houses an igniferous detonator and is closed by a perforated screwed plug secured by stabbing.

Action

Before firing

The needle is held off the detonator by the centrifugal segments which are retained in the closed position by the expanding spring ring.

During flight

On deceleration, and when the necessary rotational velocity is reached, the segments, overcoming the spring ring, swing outwards under centrifugal force clear of the needle pellet. Creep action by the needle pellet is prevented by the creep spring.

On impact or graze, the pellet moves forward, compressing the creep spring and carries the needle on to the detonator.

GERMAN GAINE, MODEL 40B
(ZÜNDLADUNG 40B)

(Fig. 10)

The gaine is used in conjunction with the AZ 38 fuze in hollow charge shell and is situated in the base of the shell cavity.

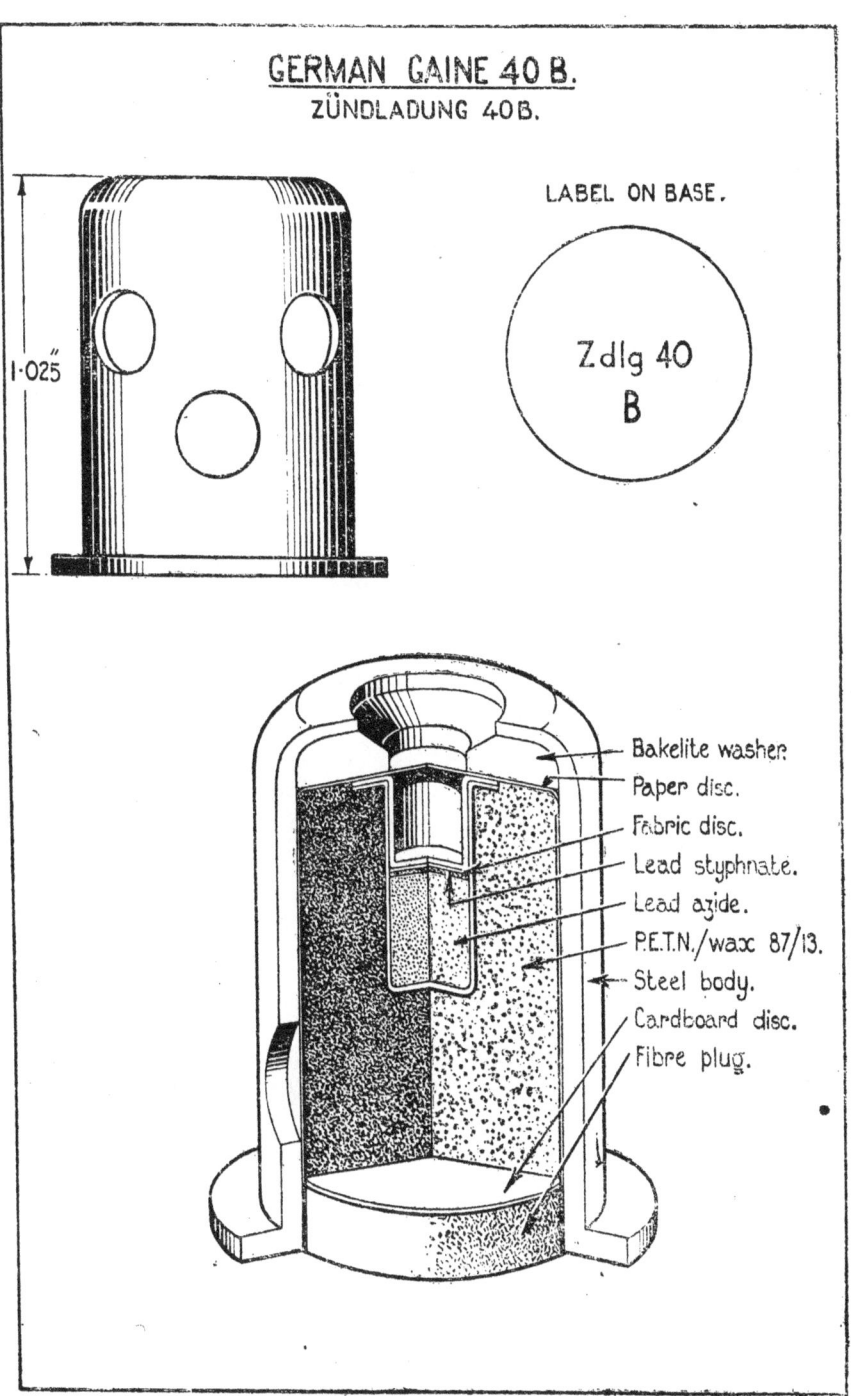

Fig. 10

The steel body is cylindrical with a flat radiussed top and a flange at the base. Two rows of four flash holes are formed in the side of the body. Externally, excepting the top of the gaine, which is not extended to form a neck, it is similar in shape and size to the Model 40 gaine described in Pamphlet No. 8. The overall length of the gaine is 1·025 inches and its diameter at the body and flange is 0·747 inches and 0·95 inches respectively. The circular aperture at the head of the gaine is 0·325 inches in diameter.

Internally at the top it is fitted with a suitably shaped bakelite washer above a transparent wafer of viscose paper, which separates the washer and filling.

The filling consists of 78·5 grains of PETN/Wax (87/13), salmon pink in colour. The detonator capsule containing approximately 4·3 grains of lead azide under a thin layer of approximately ·23 grains of lead styphnate fits into the bottom of a cavity formed in the top of the main filling. A silk fabric disc of open weave is inserted above the detonator filling and the mouth of the capsule is closed by a flanged aluminium cup.

The base of the gaine is closed by a fibre plug having a cardboard disc on its inside and, on its outside, is a stuck-on paper disc bearing the marking " Zdlg 40 B."

Action

The action is the same as the Zdlg 40.

GERMAN GAINE, MODEL 41
(ZUNDLADUNG 41)
(Fig. 11)

This gaine is used in conjunction with the AZ. 38 fuze in hollow charge shell and is situated in the base of the shell cavity.

The steel body is cylindrical in shape with a flange at its base and has a chamfered top on which is formed a tubular neck. Its overall length is 1·33 inches and its diameter at the body and neck is ·73 inch and ·32 inch respectively. There are no holes in the side of the body as in the case of the Model 40 type of gaine. The neck fits into the lower end of the central tube in the shell.

Internally, it is fitted with a suitably shaped bakelite washer at the forward end and a filling similar to that in the Model 40 gaine, but two cardboard discs are inserted in front of the screwed plug in the base.

A green label, with red bar and marked " Zdlg 41 " was stuck on the outside of the base plug.

Action

The action is the same as the Zdlg 40.

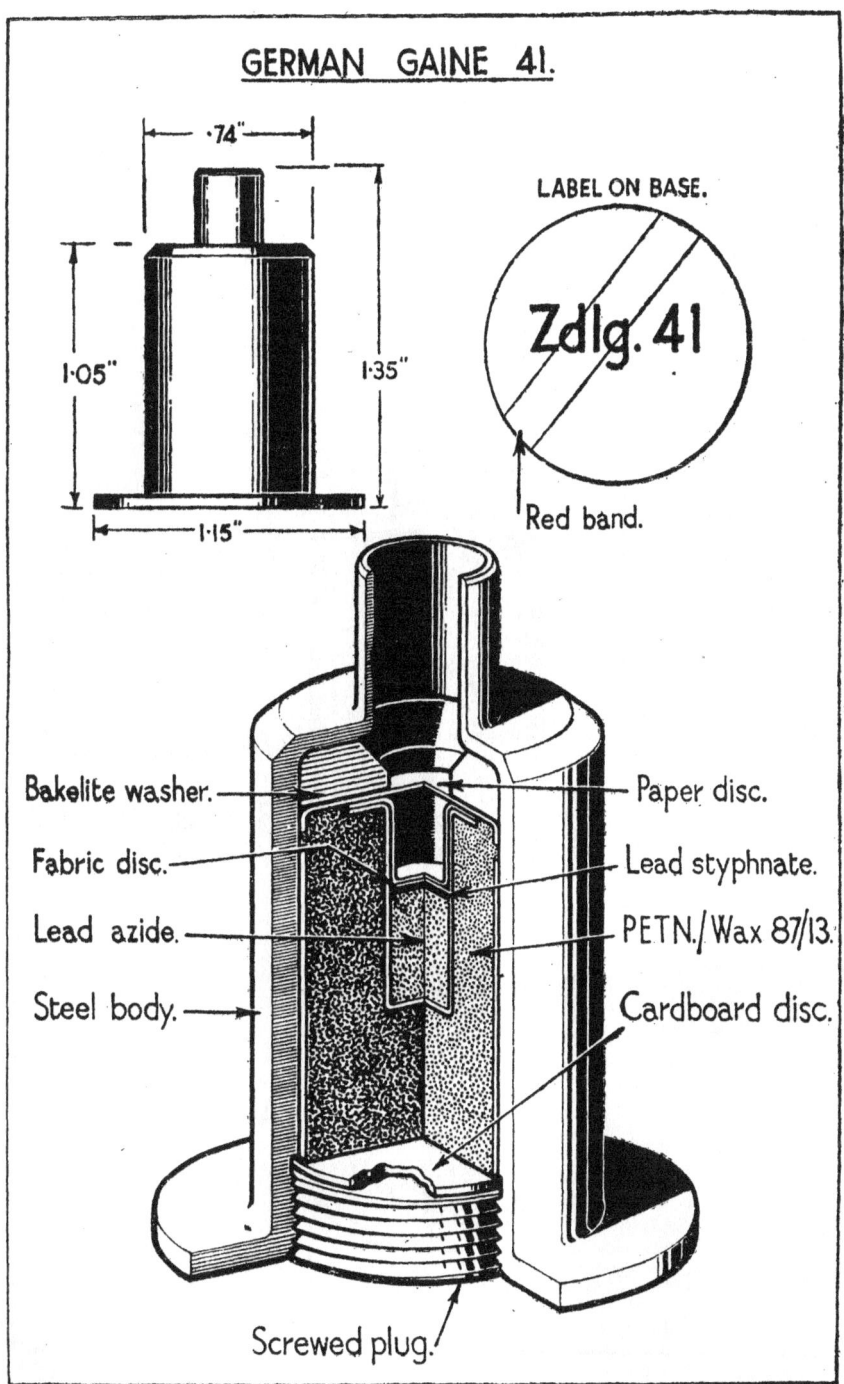

Fig. 11

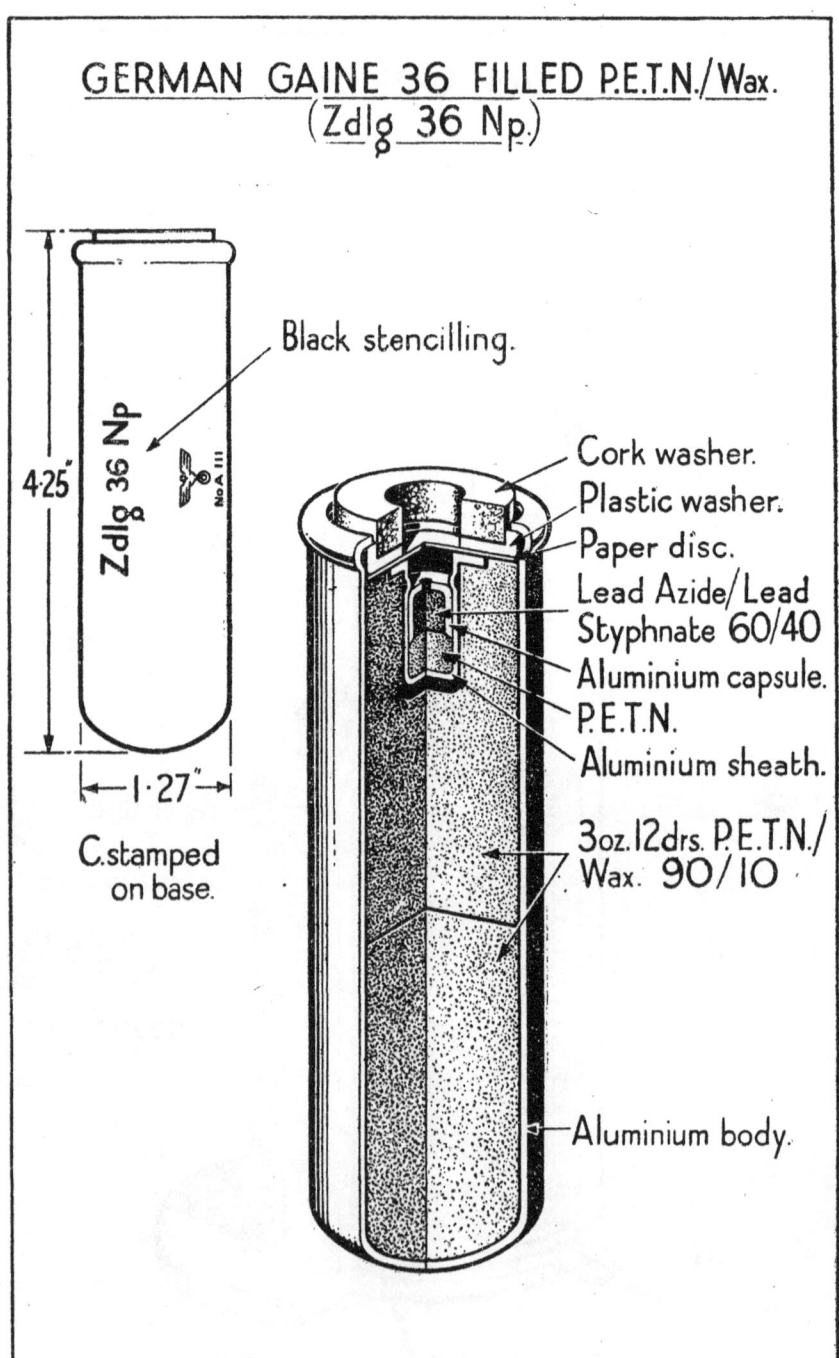

Fig. 12

GERMAN GAINE 36 FILLED P.E.T.N./WAX.
(Zdlg. 36 Np)
(Fig. 12)

Brief details of this gaine are included in the description of the 15 cm. high velocity H.E. shell in Pamphlet No. 11. The gaine has also been found in the H.E. shell for the A.A. gun 12·8 cm. Flak 40 and the H.E.B.C. shell (K. Gr. 38 (Hb)) for the 17 cm. gun (17 cm. K. Mrs. Laf.).

The overall length of the gaine is 4·36 inches and the diameter 1·27 inches. The weight complete is approximately 4 oz. 10 drs. The gaine is identified by the abbreviated designation " Zdlg. 36 Np " stencilled in black on the side of the aluminium cylindrical body. Also, unlike any of the other gaines met with, it has a convex base.

Details of the construction and the method of filling are shown in the drawing.

GERMAN HOLLOW CHARGE BOMB WITH PROJECTOR
FAUSTPATRONE 2 (PANZERFAUST 30)
(Fig. 14)

This weapon is designed for use against armour at ranges up to approximately 30 yards. It is estimated that the hollow charge will perforate 7·7 inches of homogeneous armour.

The streamlined bomb with a flat topped tapering impact cap is carried, with its flexible steel tail fins coiled, in one end of the tubular projector where it is secured by a pin. The projector, 31·5 inches long and 1·9 inches in diameter, is of the recoil-less type, held by the operator, and is fitted with a sighting arm, a bolt action firing mechanism and a propellant charge of gunpowder. With the bomb inserted in the projector the weight is approximately 11 lb. 6 oz., and the overall length 41 inches. The impact cap of the bomb is painted buff colour; the rustproofed body is blue-black in colour; the tail portion, except the fins, which are unpainted, is painted buff colour. The projector is painted buff colour and has stencilled on it, in red, an arrow pointing to the rear end with the words " Achtung Feuerstrahl! " This wording is, in effect, a warning to beware of propellant gases issuing from the rear end of the projector.

Bomb (Fig. 13)

The hollow charge bomb is of light construction with an overall length of 19·5 inches and maximum diameter of 5·6 inches. The bomb examined was deficient of the gaine and fuze and in this condition weighed 6 lb. 7½ oz. The steel body tapers towards the base, where it is rounded and has a central hole and is fitted with an extension tube which normally contains a gaine and a base fuze.

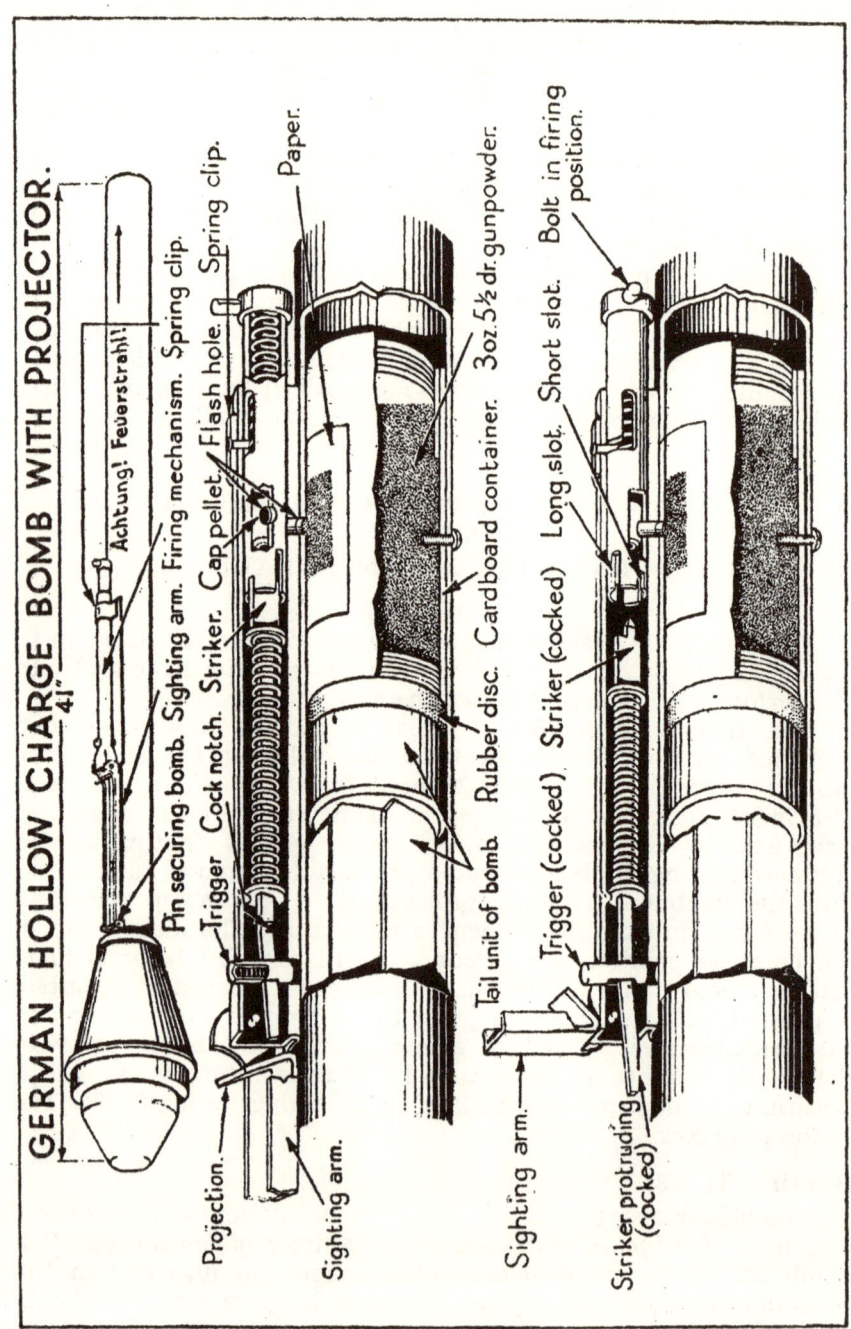

Fig. 14

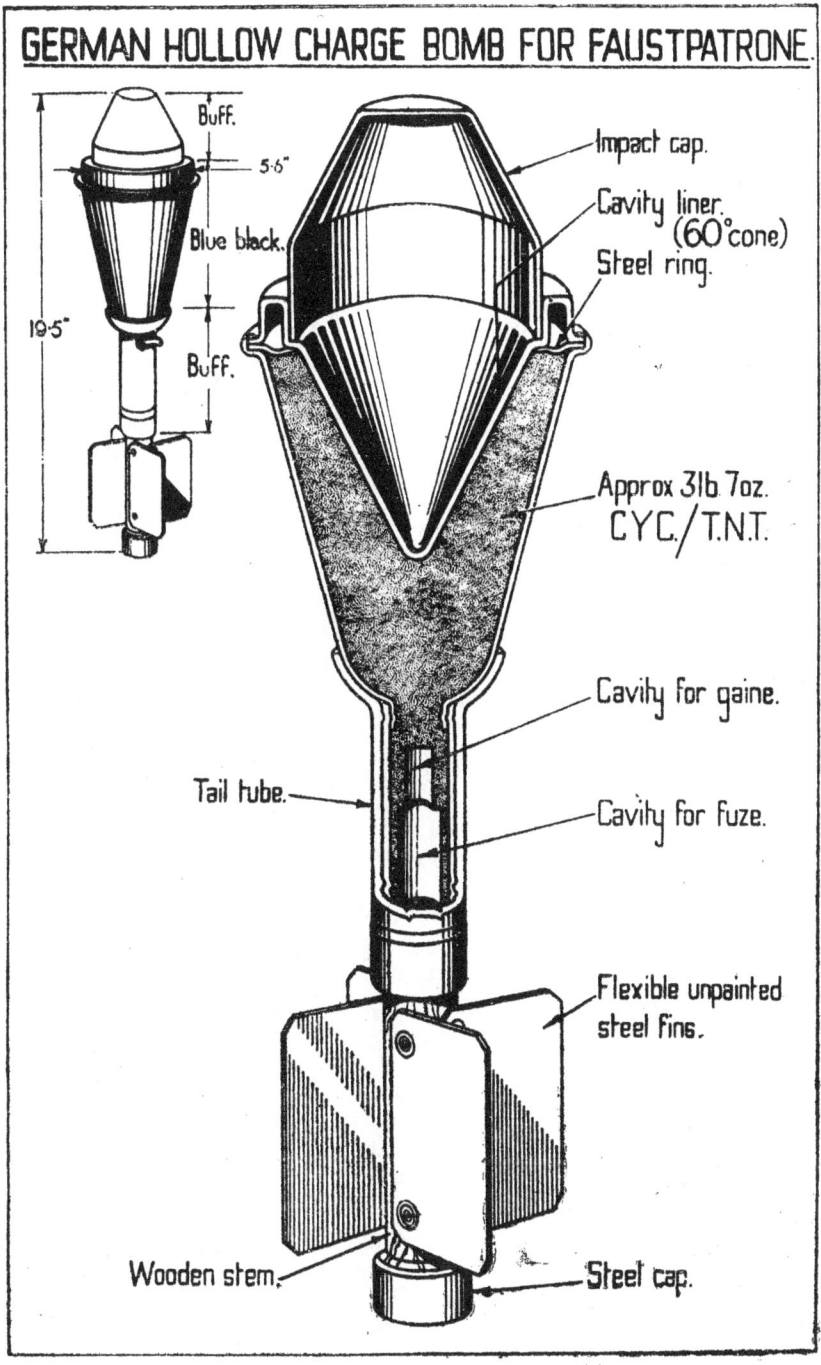

Fig. 13

The lower end of the extension tube has a pressed screwthread to engage projections in the tail tube for the attachment of the tail to the bomb. The body contains a 3 lb. 7 oz. hollow charge of cyclonite and T.N.T. with a cavity liner in the form of a 60 degree cone. The top of the liner is shaped to form a cylinder for attachment to the body and for the assembly of the impact cap which is sprung into the cylinder. A steel ring is inserted between the top of the filling and the cylindrical portion of the liner. Cavities for a base fuze and a gaine are formed in the lower part of the H.E. filling within the extension tube.

The tail unit of the bomb consists of a steel tube with a wooden stem carrying four flexible steel fins. The tail tube is shaped at the front end to fit over the rounded base of the bomb body, and on the outside is fitted with a small bracket to receive a pin which secures the bomb to the sighting arm on the projector for transport. The wooden stem to which the fins are riveted has a steel base cap.

Fuze (F.P.Z.8001)

Described as a separate item in this pamphlet.

Gaine

This is a small P.E.T.N./Wax gaine in an aluminium case, the dimensions of which are identical with the " K1 Zdlg. 34 Np " described in Pamphlet No. 11.

Projector (Fig. 14)

The steel tubular projector with the propellant charge fitted inside weighs 4 lb. 14¼ oz. The tube has an internal diameter of 1·72 inches and is open at both ends. At the forward end of the tube a portion is cut away to accommodate the bracket on the bomb. The propellant charge is situated at 9·8 inches from the forward end of the projector and extends to within 17·2 inches from the rear end. The charge consists of 3 oz. 5½ drams of granular gunpowder in a cylindrical cardboard container. The container is 4·1 inches long and is secured in position by a screw inserted in the side of the projector. The ends of the container are made up of a number of cardboard discs and a rectangular aperture in the side is closed by a piece of waxed transparent paper. The aperture is positioned to correspond to a flash hole in the side of the wall of the projector communicating with the firing mechanism. A circular disc of rubber-like material is inserted above the container to act as a cushion for the tail of the bomb.

The firing mechanism is contained in a casing in the form of a short tube attached to the top of the projector. A sighting arm is hinged to the forward end of the casing. The arm is of channel section steel with a sighting aperture near its centre and two notches at the forward end to engage the pin, which secures the bomb to the projector for transport. At the hinged end of the arm a projection is formed which prevents the firing mechanism being cocked when the arm is folded down in the position for transport.

The firing mechanism consists of a striker with spring and a trigger in the form a spring loaded plunger and a cocking bolt which acts against the firing end of the striker and carries a percussion cap.

The striker is rectangular in section and is in the form of a rod enlarged at the pointed end and inclined at the opposite end. A cocking notch is formed at the commencement of the incline. The spiral spring is assembled over the striker between two circular end plates. The plates each have a rectangular hole for assembly on the striker. The plate at the forward end is supported by an indentation in the casing and that at the rear end by the enlarged head of the striker.

The trigger is in the form of an inverted steel cup with rectangular holes in the side through which the inclined part of the striker passes and containing a spiral spring. The spring is held in compression between the edge of the striker and the closed end of the cup and tends to move the trigger outwards through a hole near the forward end of the casing.

The cocking bolt enters the rear end of the casing and has a small handle screwed into it for manipulation. The bolt is retained in the casing by a pin fitted to a spring clip which fits around the casing. The pin is inserted through a hole in the casing and enters a slot in the tubular body of the bolt. The slot and pin limit the movement of the bolt and permit the bolt to be turned through 90 degrees. The forward end of the bolt has two short slots diametrically opposite and two long slots similarly placed. The short slots correspond to the position of the bolt handle and engage the enlarged head of the striker when the striker is in the uncocked position. The long slots are situated at 90 degrees from the short. The bolt contains a spiral spring held between its rear end and the pin of the spring clip, a small steel end plate being fitted at its forward end to bear against the pin. The percussion cap is carried in a steel cylindrical pellet inside the forward end of the bolt. A flash hole is formed in the side of the pellet corresponding to a similar hole in the bolt. The pellet is secured to the bolt by the metal around the hole in the bolt being turned into the hole in the pellet.

Action

The pin securing the bomb to the projectile is removed before firing. The sighting arm is opened to a position at right angles to the projector and the weapon is laid by aligning the rim on the cylindrical part of the bomb with the target through the aperture in the sighting arm; the rim thus acting as a foresight. The weapon may be used in the standing, kneeling or lying position, but care must be taken to place the weapon so that nothing obstructs the path of the propellant gases to the rear. According to a German training pamphlet, a jet of flame approximately 6 feet in length shoots out of the rear end of the projector on firing. The firer must wear a steel helmet and immediately after firing must take cover to avoid being hit by splinters. The danger area behind the projector is approximately 30 feet.

The firing mechanism is cocked by pushing the bolt forward with the handle in the vertical position. The striker, engaged in the short slots in the front end of the bolt is moved forward with the bolt and the striker and bolt springs compressed. The inclined forward end of the striker emerges from the front end of the casing and, when the notch in the striker reaches the trigger, the spring in the trigger raises the cylindrical body so that the notch is engaged. The striker is then in the cocked position with the trigger and front end protruding from the casing. Unless the sighting arm is raised the projection near the hinged end of the arm obstructs the forward movement of the striker and prevents the mechanism being cocked. After cocking, the bolt is released and is returned to the withdrawn position by its spring. With the bolt in this position and the handle vertical, the short slots at its front end are still in alignment with the striker head. This is apparently intended as a safe setting for the cocked mechanism, as the length of the slots would prevent the striker reaching the percussion cap in the bolt. Also the flash holes in the bolt and body are not coincident. When the bolt is turned through 90 degrees in a counter clockwise direction the long slots in its front end are brought into alignment with the striker head and the flash holes coincide. When the trigger is pressed in, it is disengaged from the striker notch and the striker is driven back by its spring to pierce the cap. The flash from the cap enters the projector through the flash holes and ignites the propellant charge which projects the bomb. The escape of some of the propellant gases through the rear end of the projector prevents the firer being subjected to a recoil force.

GERMAN 8 cm. MORTAR H.E. BOMBS 38 umg ·AND 39 umg

It has now been established that the H.E. bombs Wgr 38 and Wgr 39 (described in Pamphlet No. 11) designed to burst in the air after impact, have been converted to normal H.E. bombs. The converted bombs are designated " Wgr. 38 umg." and " Wgr. 39 umg." and are stencilled " 38 umg." and " 39 umg." respectively. These bombs range the same as the standard H.E. bomb " Wgr. 34 " described in Pamphlet No. 6, page 9.

GERMAN 7.3 cm. PROPAGANDA ROCKET

(Fig. 15)

This rocket is designed for dispersing propaganda leaflets, but the projector from which it is fired has not been identified. The overall length of the complete round is 16·1 inches, its maximum external diameter 2·87 inches and, including leaflets (weighing 8 ozs.), weighs 7 lb. 2 oz. It is fired by percussion and fitted with base venting venturis, half of which are inclined to rotate the rocket in flight. " Pr. G. 41 " is painted on the lid of the metal box container which holds 4 complete rounds.

The rocket consists mainly of :—

Nose cap.
Head tube with split case to accommodate propaganda leaflets.
Ejector with fixed delay unit.
Connecting collar.
Tail tube containing propellant charge and ignition system.
Venturi block with percussion cap.

The nose cap is of plastic material weighing 4 oz. shaped to a low crh. and slightly reduced in diameter near the base end, which fits into the head tube. The cap is easily detached from the tube for insertion of reading matter.

The head tube is a steel cylinder 7·48 inches in length and open at both ends. The external and internal diameters are 2·56 inches and 2·36 inches respectively. Internally, at the forward end, the diameter is slightly increased to receive the nose cap which fits against a shoulder. The rear end of the tube is screwthreaded externally to enable the head and tail tube to be united by means of a connecting collar.

The leaflets are rolled and contained in a split case which is a sliding fit in the head tube. The split case is held between the base of the nose cap at the forward end and two millboard discs above a plastic disc at the base. The discs act as a piston in ejecting the leaflets. A light V-shaped spring held in compression fits into the centre of the roll of leaflets.

The ejector unit is housed immediately below the plastic disc ; it consists of a short brass tube with an internal diaphragm separating a top and bottom chamber. The chambers are connected by a communicating hole in the diaphragm displaced from the centre.

The forward chamber contains the ejection charge consisting of approximately 1·13 drams of gunpowder and is closed by a millboard disc. The rear chamber is screwthreaded internally and houses a delay holder. The delay holder is screwthreaded externally and, in its upper surface, provided with a circular groove filled with a delay composition of slow match. The delay composition is in communication with the ejector charge and the central flash hole in the base of the holder. No box washer is used to cover the delay composition and no escape is provided for the products of combustion. The time of burning is not adjustable. The flash hole in the base of the holder is screw-threaded to receive a short tubular adapter, filled with gunpowder, which also screws into a steel disc fitted into the head of the tail tube.

The connecting collar unites the head and tail units. It has an internal flange formed near its centre and is screwthreaded internally to receive the head and tail tubes.

The tail tube is 6·35 inches in length and of similar wall thickness and bore to that of the head ; it is screwthreaded internally at each end and, at its forward end, provided with an internal flange spun over to form a groove in its under side. At the venturi end, the internal surface is chamfered.

A steel disc 0·4 inch thick with a stepped upper surface; fits into the head of the tail tube so that the step bears against the internal flange of the tube. The disc is bored centrally and screwthreaded to receive the adapter.

Below the steel disc is an igniter support. The support consists of a washer with a ring, ꞁ shaped in section, to accommodate the igniter centrally below the adapter, and three small pieces of spring steel equally spaced near the periphery, with their lower edges bent inwards, all welded to the underside. The latter serve as distance pieces between the propellant charge and the igniter.

The igniter consists of a pressed pellet of gunpowder contained in a circular flat aluminium capsule 1·17 inches in diameter, with its open side facing the propellant.

The propellant charge consists of a single cylindrical stick of propellant with a central axial hole surrounded by eight annular holes throughout its length. The stick weighs 15¾ oz., is 5·25 inches long and 2·25 inches in diameter. Three longitudinal ribs are formed on the propellant when extruded from press. It is presumed that they are designed to fix the charge during transport and centralize it for ease of ignition. There are no metal grids. The main stick of propellant is held off the venturis by a small washer of propellant approximately 0·16 inch thick and 1·06 inches diameter of the same composition as the main charge, cemented to the base of the propellant.

A provisional analysis of the propellant is as follows, nitrocellulose 60·6 per cent., diglycoldinitrate 32·9 per cent., akardite 0·8 per cent., graphite 0·2 per cent., ash (carbonated) 0·7 per cent., and undetermined matter 4·8 per cent.

No auxiliary materials are employed to transmit the flash from the percussion cap to the igniter.

GERMAN 7·3 cm. PROPAGANDA ROCKET.

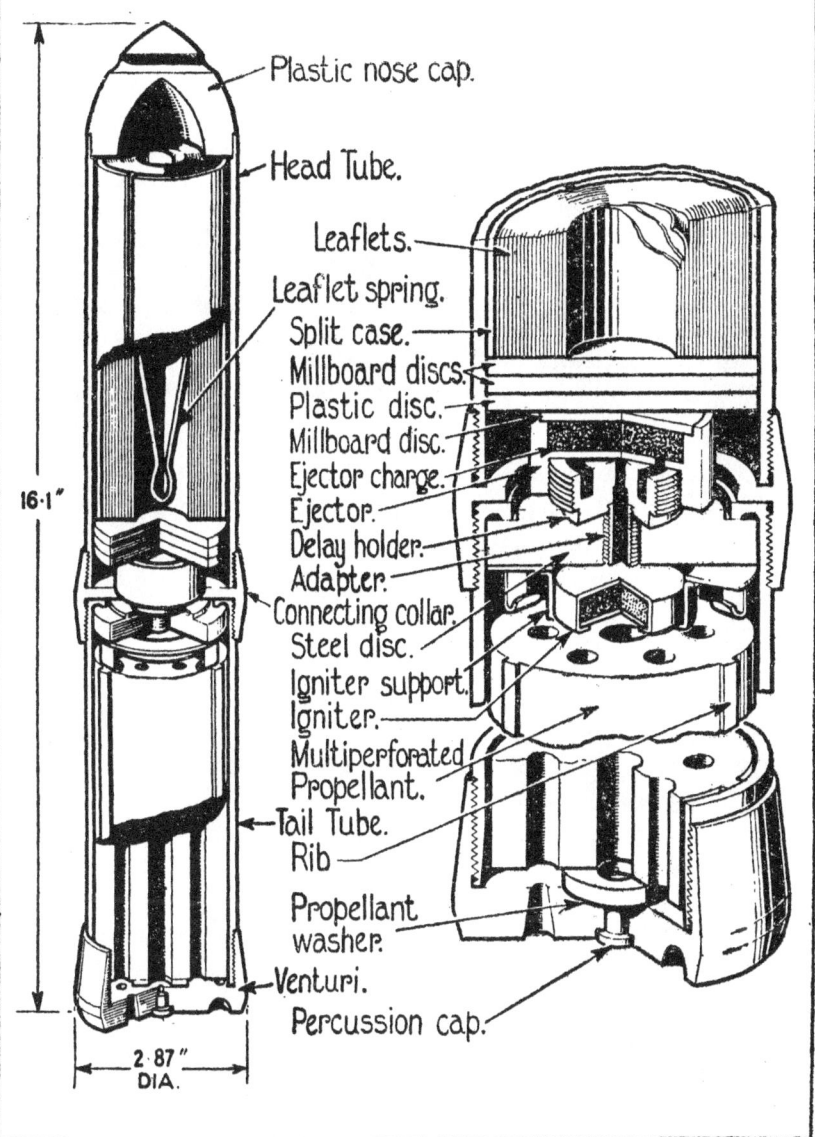

Fig. 15

The venturi block is screwthreaded internally for attachment to the tail tube and is chamfered at the base. It is provided with 14 coned venturis arranged in two concentric circles of seven in each, and a central hole bored in three diameters to house a percussion cap. The outer ring of venturis have a throat diameter of 0·118 inch and are inclined about 30 degrees, while those forming the inner ring have a throat diameter of 0·138 inch and are not inclined. The full angles of the emergent cones are about 30 degrees. No attempt has been made to seal the venturi end of the tail tube.

Action

On the percussion cap being struck, the flash passes through the central hole in the propellant to the igniter which ensures complete ignition of the propellant charge. The igniter also ignites the priming composition in the adapter, which in turn ignites the delay composition. Pressure set up inside the rocket and escaping through the venturis propels the rocket forward, the inclined outer ring of venturis causing the rocket to rotate.

The delay composition burns until such time as it ignites the ejector charge. Gas pressure from the ejector charge forces the discs forward to expel the contents of the head tube. As the split case emerges it falls apart allowing the spring to scatter the leaflets.

GERMAN 8·8 cm. ANTI-TANK HOLLOW CHARGE ROCKET PROJECTILE
(8.8 cm. R. Pz. B. Gr. 4322)

(Figs 16 and 17)

This self-propelled streamline hollow charge rocket projectile is fired electrically from the German 8·8 cm. anti-tank rocket projector (Raketen Panzerbüchse 54). The rocket is loaded into the rear end of the projector and retained in position by a retaining catch and spring loaded plunger. A jet of flame shoots out of the rear end of the projector when the rocket is fired.

The rocket consists of :—

  Hollow charge bomb filled Cyclonite/TNT. (60/40) ;
  Fuze A.Z. 5095 ;
  Gaine Kl. Zdlg 24. Np. ;
  Tail unit with propellant charge ;
  Electric fuze and igniter.

The overall length of the complete round is 25½ inches and its weight 7 lb. 4 oz. The propellant is contained in the tail tube, and in flight the rocket is stabilized by fins and not by rotation. The rocket is painted a deep olive green and the bomb is stencilled in black.

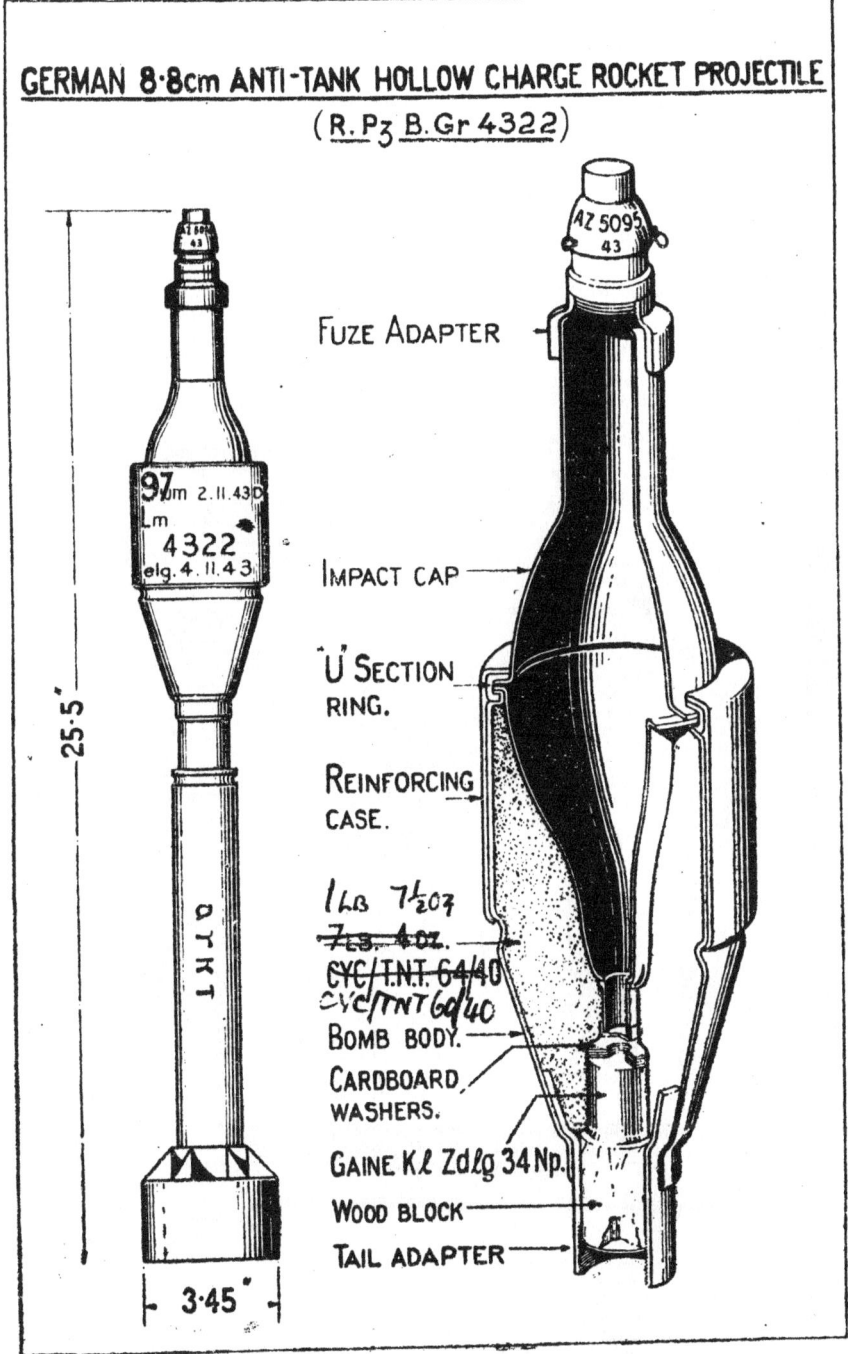

Fig. 16

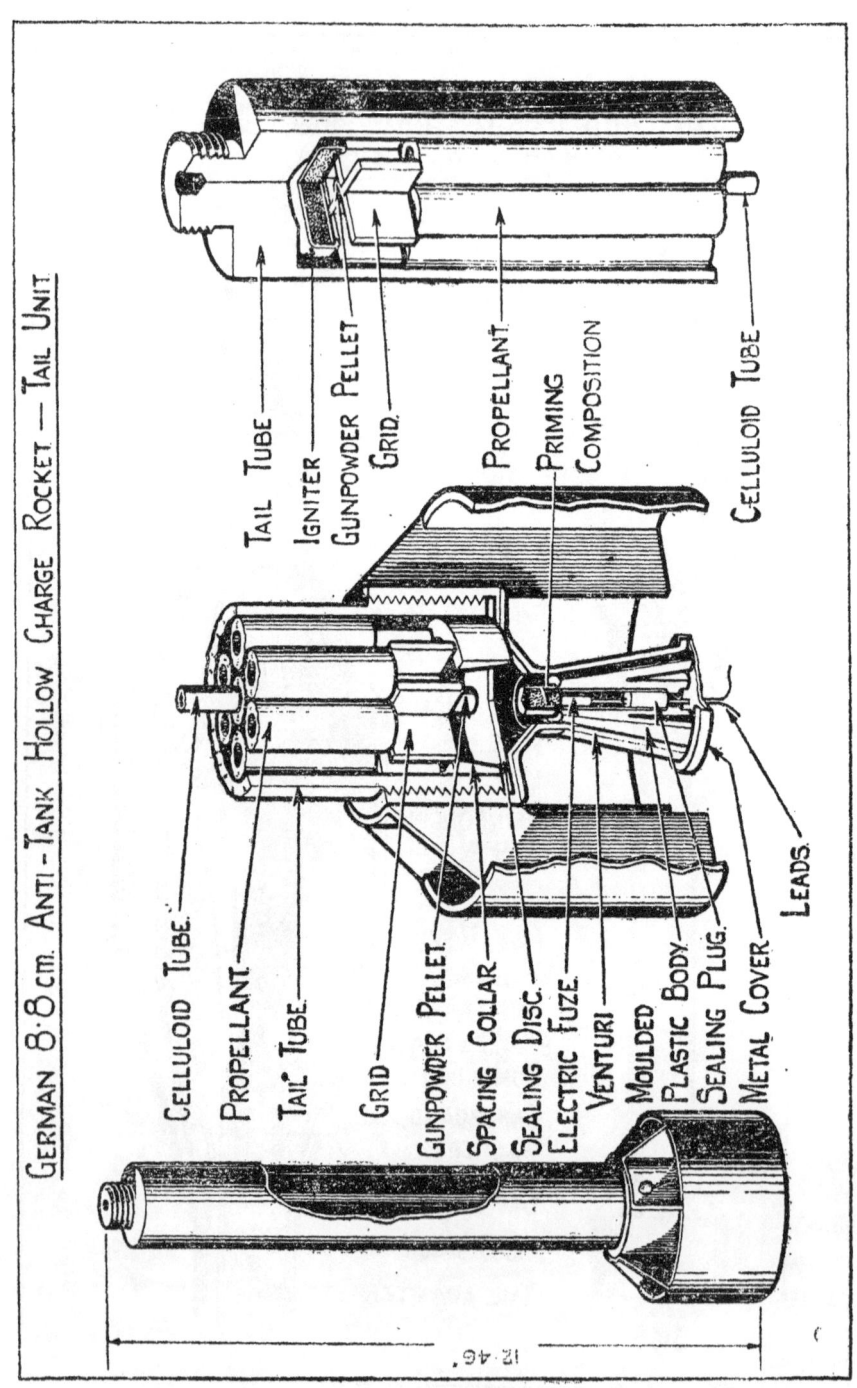

Fig. 17

Bomb

The hollow charge bomb is of light construction, streamlined, and has a maximum diameter of 3·5 inches. The length of the bomb with fuze is approximately 13·75 inches. It consists essentially of two main parts, the bomb body containing the filling and the impact cap.

The bomb body is of thin steel, the upper half is cylindrical in shape and the lower half tapers towards its base. A tail adapter, fitted internally, and shaped to the bomb body, protrudes from the base; it is screwthreaded internally to receive the upper end of the tail unit. A fold at the forward end of the body enables the flanged base of the impact cap to be secured to it by a " U " section ring. The " U " section ring is retained in position by a thin tubular reinforcing case surrounding the bomb body at its maximum diameter. The reinforcing case has an internal flange at its forward end which fits over the " U " section ring, and has its rear edge secured by turning it into a circumferential groove formed in the body.

The body contains a 1 lb. 7½ oz. hollow charge of cyclonite and T.N.T., with a cavity liner approximately 3·7 inches deep, shaped like the stem half of a pear and terminating in a central cylindrical chamber of two diameters formed in the base of the filling. The lower portion of this chamber is larger in diameter and length than the upper, and accommodates the gaine under four cardboard washers. The base of the bomb body is closed by a cylindrical wooden block.

The impact cap, approximately 4·75 inches in length, is coned from its base to about half its length and terminates into a cylindrical head approximately 1·3 inches in diameter. The end of the head fits into an adapter, the forward end of which is reduced in diameter and screwthreaded internally to receive the fuze.

Fuze

The fuze A.Z.5095 is described as a separate item in this pamphlet.

Gaine

The gaine, Kl. Zdlg. 34 Np. is described in Pamphlet No. 11, page 30.

Tail Unit (Fig. 16)

The tail unit, including the fins, has an overall length of 12·46 inches and weighs approximately 2 lb. 15½ oz. It consists principally of the tail tube, venturi and fin assembly, propellant charge, two grids, steel collar, igniter system and an electric fuze.

The metal tail tube is cylindrical in shape, apparently solid drawn and turned on the outside, with an external and internal diameter of approximately 1·57 inches and 1·38 inches respectively, and an overall length of 10·69 inches. The forward end is closed and formed with an external screwed boss of smaller diameter which, in assembling, screws into the adapter at the base of the bomb. The rear end is screwthreaded externally to receive the venturi and fin assembly.

The venturi and fin assembly are an integral unit. The forward end of the venturi is screwthreaded internally to receive the rear end of the tail tube, which screws in on to a shoulder. Immediately in rear of the shoulder, the venturi is sharply coned to a smaller diameter, and then opens to an angle of approximately 10 degrees to the centre line. The latter expanding portion is welded to the entry at the throat, which is 0·52 inch in diameter. The fin assembly consists of a drum with six radial fins. The fins are made in pairs and welded to the drum.

The propellant charge weighs 6 oz. 2·8 drs. and consists of seven tubular sticks of propellant each 7·9 inches long, 0·45 inch and 0·20 inch external and internal diameters respectively. Preliminary chemical analysis has shown that the propellant contains approximately 64·7 per cent. nitrocellulose and 34·2 per cent. diglycoldinitrate, and 1·1 per cent. undetermined matter including stabilizer. A central stick of propellant is surrounded by the other six and held between two grids, one at each end of the propellant. The whole is in turn held between an igniter, forming part of the ignition system, in the forward end of the tail tube and a coned steel spacing collar at the rear.

The grids are identical and each consists of four rectangular steel plates with two slots cut into one side to half the width of the plate. Four plates are assembled to form a cross-like grid with nine squares of equal size. The size of the squares in the grid are such as to prevent the charge slipping through them. The base grid is supported by a steel spacing collar, coned internally, with its larger diameter to the rear.

The igniter system consists of a fuze with priming composition, fitting into the venturi mouth, an igniter at the forward end of the motor tail, and a central celluloid tube with gunpowder pellets at each end.

The igniter is in the form of a pressed pellet of black powder in a flat circular aluminium capsule situated in the forward end of the tail tube. The mouth of the capsule faces towards the propellant and the composition is covered by a thin disc of cellophane-like material.

The celluloid tubing is in two lengths, and is a sliding fit in the central stick of propellant. It contains no quickmatch. The forward end of the tube contains a cylindrical perforated pellet of gunpowder accommodated in the centre square of the forward grid. The rear end of the tube passes through the rear grid, and the cylindrical perforated pellet in this end is accommodated in the steel spacing collar.

Electric Fuze

The electric fuze consists of a metal tube, a bridge wire, two leads, and priming composition.

The front end of the tube is closed by a metal disc which is spun over the end of the tube. The tube contains the fuze bridge wire at the end of the two leads. The leads and bridge wire are held in position by a plug of polyvingl-chloride through which the leads pass, the metal tube being crimped at this point to prevent movement of either fuze wire or plug. The tube is filled with finely granulated reddish brown powder which surrounds the bridge wire. The powder consists of an oxide of lead, approximating in composition to red lead, bound together with nitrocellulose. The nitrocellulose constitutes less than 5 per cent. of the composition.

The resistance of the bridge wire is approximately 10 ohms. The power, provided by the small projector generator, has a peak voltage of approximately 1·2 volts. It appears that the high sensitiveness of the fuze is due to the priming composition rather than the bridge wire.

The fuze is housed in a conical shape body of moulded plastic. The body is bored centrally in three diameters to make an internal flange near the base. The long chamber formed in front of the flange houses the tube portion of the fuze. The conical body is cemented into the divergent portion of the venturi which is closed by a metal cover through which the leads pass. The end of one lead is soldered to the fin assembly.

Action

Before loading, the safety pin is withdrawn from the nose fuze.

Tail Unit

On pressing the projector trigger, an electric current is generated which heats the bridge wire of the base fuze and fires the priming composition. The flash from the priming composition in turn ignites the gunpowder at each end of the celluloid tube and the igniter. The igniter ensures the complete ignition of the propellant charge. Pressure set up inside the rocket and escaping through the venturi propels the rocket forward.

Bomb

On impact, the impact cap is shattered and the fuze functions. The flash from the fuze passes through the cavity in the bomb filling to the gaine. This results in the detonation of the gaine, which in turn brings about the detonation of the bursting charge. The " hollow " in the head of the bursting charge has the effect of concentrating the detonation forward on to the plate struck.

GERMAN 15 cm. H.E. ROCKET
(15 cm. Wurfgranate 41 Spr.)
(Fig. 18)

This is a self-propelled, circumferential venting H.E. rocket filled T.N.T. The shell is in rear of the propelling unit. It is fired electrically and projected from either the 15 cm. Panzerwerfer 42 or the 15 cm. Nebelwerfer 41 projector.

The general shape of the rocket and its external markings are shown in Fig. 18. The complete round weighs, approximately 76 lb. 10 oz., and its overall length is 36·4 inches. It is painted either a deep olive green or light grey green and stencilled in white on the propelling unit and in black on the shell.

The complete round consists of the following principal components :—

Shell
    Shell filled T.N.T. ;
    Base fuze Bd. Z. DOV. ;
    Gaine Gr. Zdlg. C/98 Np. ;
    Motor unit ;
    Electric ignition fuze.

The body is of malleable iron, cylindrical in shape with a radiused base. The overall length, including the obturator, is 10·4 inches, and the external diameter 4·97 inches. The wall of the body is approximately 0·4 inch thick. The base is bored centrally and screwthreaded to receive a fuze adapter to which the steel gaine container is attached by indenting the mouth of the latter to an internal cannelure in the former. The forward end of the body is screwthreaded internally to receive the base of the venturi block which screws on to the flange of an inverted cup-shaped cast iron obturator, accommodated in a recess in the base of the venturi block.

The bursting charge, weighing approximately 5 lb. 10 oz., consists of two pressed pellets of T.N.T. completely enclosed in a shaped cardboard carton with a cavity to accommodate the gaine at the rear end and reduced in diameter at the forward end to fit into the obturator. The carton is cemented to the body by a magnesium oxychloride composition.

Base Fuze—Bdz. DOV.

This fuze is described as a separate item in this pamphlet.

Gaine—Gr Zdlg C/98 Np.

A description of this gaine is given in Pamphlet No. 4, page 33, and illustrated in Fig. 22. .

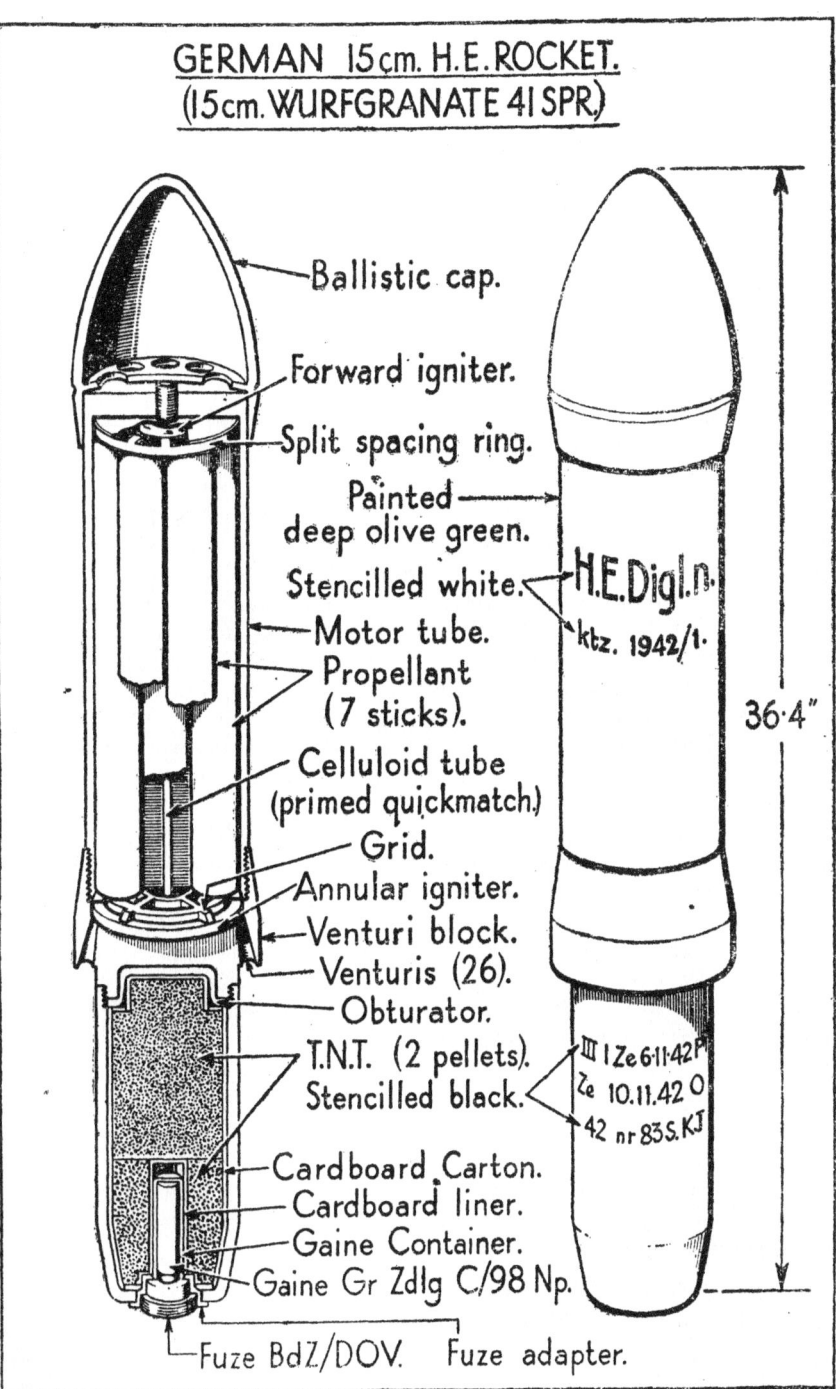

Fig. 18

Motor Unit

The motor unit with venturi block weighs approximately 50 lb. 11 oz., and has an overall length of 18·9 inches. It consists mainly of a tube, ballistic cap, venturi block, propellant charge, grid, spacing ring and ignition system.

The tube is cylindrical with a closed end and appears to be a solid forging, machined inside and out to a diameter of 5·05 inches and 5·5 inches respectively, and to an internal depth of 18 inches. It weighs 24 lb. 13 oz. The closed end is bored centrally and screw-threaded for the attachment of the ballistic cap. The open end is chamfered internally and screwthreaded externally to receive the venturi block.

The ballistic cap is of thin pressed steel and provided with a perforated internal diaphragm with a central screw to enable the cap to be secured to the tube.

The venturi block, which unites the motor unit and shell, weighs 10 lb. 14 oz. The top is cup-shaped and screwthreaded internally to a depth of 1·75 inches to receive the tube; externally it is chamfered from the skirt towards the top. The base is provided with a screwed boss for the attachment of the shell unit and is recessed to accommodate the shell obturator. Twenty-six venturis, equally spaced to form an annular ring, are bored in the skirt near the periphery. Each venturi has a throat diameter of 0·22 inch and is inclined at 14 degrees to rotate and stabilize the rocket in flight. The venturis are sealed by a thin aluminium disc inserted on the inside.

The propellant, as indicated by the stencilling " Digl " on the tube, is of the double-base type consisting of nitrocellulose 61·1 per cent., diethylene—glycol—dinitrate 33·3 per cent., stabilizer (probably akardite) 2·1 per cent., graphite 0·24 per cent., ash (carbonated) 0·75 per cent., and undetermined matter 2·51 per cent. The charge weighs 13 lb. 1 oz. and is in the form of six tubular sticks surrounding a seventh. Each stick is 16 inches long, 1·65 inches and 0·25 inches external and internal diameters respectively. The charge is held between a grid at the base and a spacing ring at the head.

The grid, which weighs 1 lb. 3 oz., is a cast iron ring, 4·4 inch external circumference on a web of six spokes radiating from a central ring. It is centred by the chamfered interior of the tube, which suits the chamfered end of the spokes. Longitudinal movement of the grid is prevented by means of a felt disc which is compressed when the venturi is screwed to the tube.

The spacing ring at the head is a split ring with ten pairs of lugs bent inwards to form a " U " shape in section.

The ignition system comprises a rear and forward igniter and a length of primed celluloid tubing.

The rear igniter consists of an annular transparent plastic ring containing approximately 1 oz. 7 dr. of black powder, and a silk flat bag containing approximately 3 dr. of gunpowder. The ring is housed immediately inside the venturis. Stuck to the inside of the ring is an opaque plastic flange with a felt disc cemented to it. The felt bag is sewn to the centre of the felt disc.

The forward igniter consists of 3 oz. of gunpowder in a circular cardboard box which is housed in a metal cup with a wide flange. The flange is held against the head of the tube by the spacing ring so that the igniter is held centrally within the ring and facing the central stick of propellant. The igniter ensures complete ignition of the propellant.

The celluloid tubing is housed in the central stick of propellant. It contains quickmatch and is closed at each end by a small gunpowder pellet.

Electric Ignition Fuze

The electric ignition fuze is described as a separate item in this pamphlet. It is inserted in one of the venturis.

Action

Motor Unit

An electrical current generated from a remote control fires the electric fuze, thereby producing a flash which in turn ignites the rear igniter, the primed celluloid tubing, the forward igniter and the propellant charge. Pressure set up by the gases and escaping through the venturis propels the rocket forward. The venturis, being inclined cause the rocket to rotate and so stabilize it in flight.

GERMAN 21 CM. H.E.B.C. ROCKET

21 cm. Wurfgranate 42 Spr.

(Figs. 19, 20 and 21)

This is a self-propelled base venting H.E. rocket stabilized by rotation and fired electrically from either the 21 cm. Nebelwerfer 42 projector (a two-wheeled field carriage with five barrels) or the 21 cm. airborne projector beneath each wing of the ME 109G–6 (Trop) and FW190. When fired in the former the nose fuzes le 1gr. Z. 23 nA., or Hbgr. Z. 35 K., are used whilst in the airborne projector the clockwork fuze Zt. Z. S/30 is employed. The complete round weighs approximately 242 lb. and its overall length is 49·5 inches. The centre of gravity is 18 inches from the base. The exterior of the rocket is painted deep olive green and stencilled in black. External markings on the round are shown in Fig. 19.

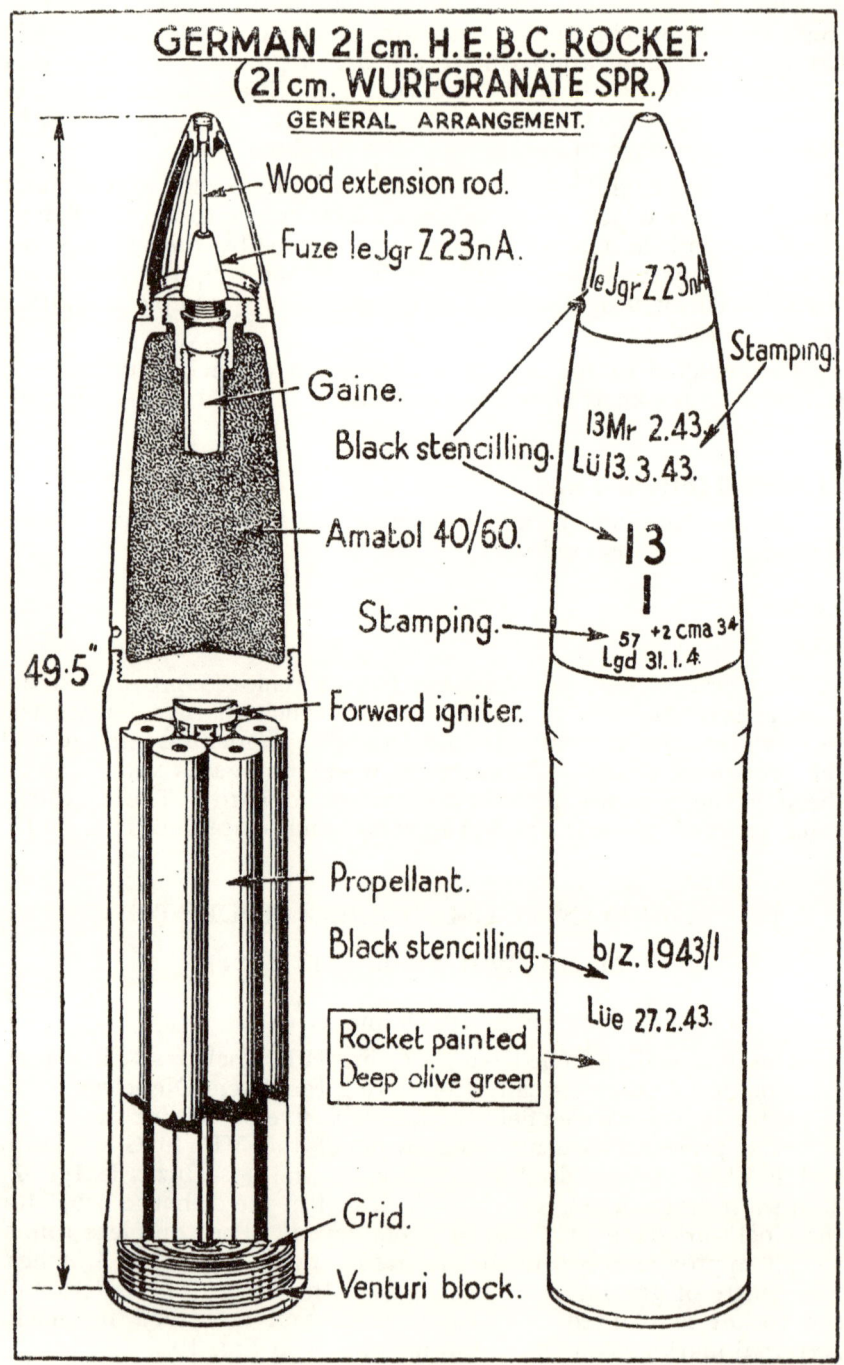

Fig. 19

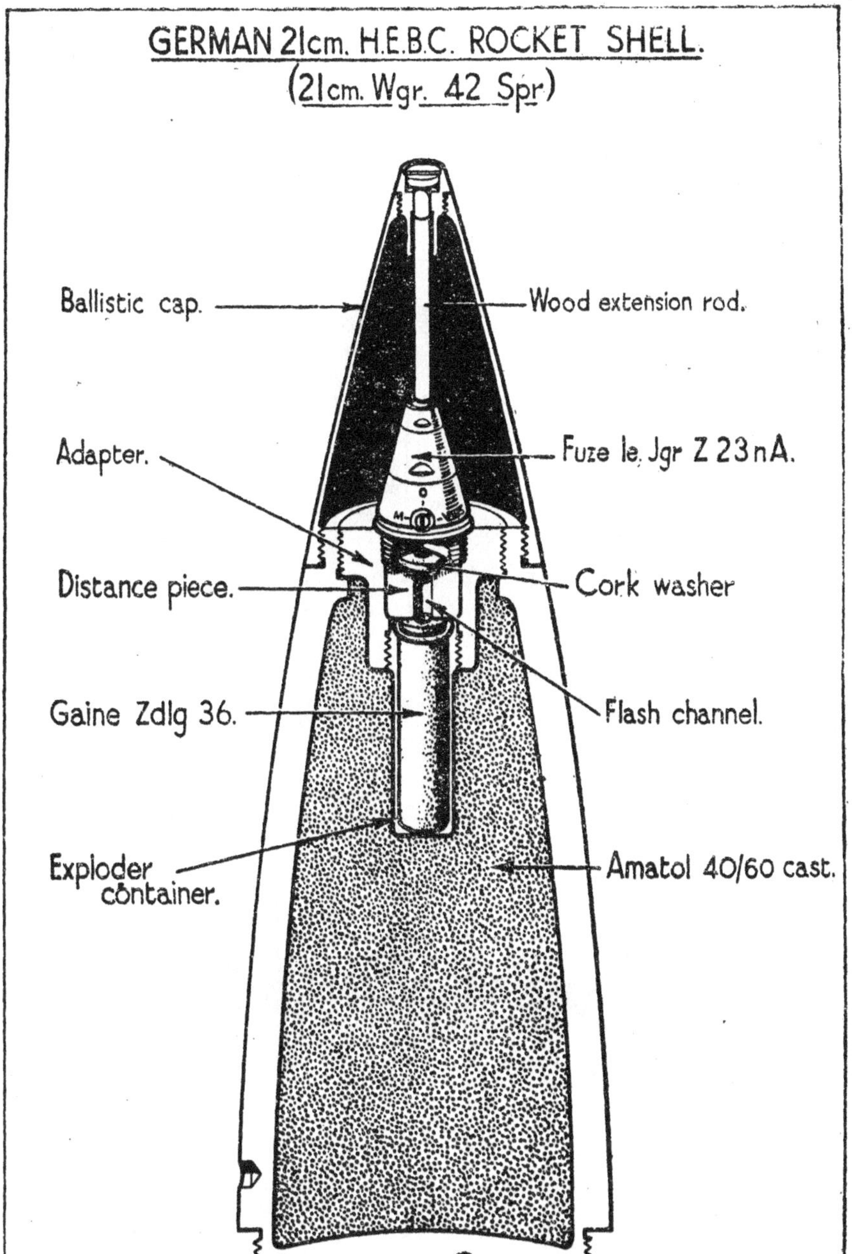
GERMAN 21cm. H.E.B.C. ROCKET SHELL.
(21cm. Wgr. 42 Spr)
Ballistic cap.
Wood extension rod.
Adapter.
Fuze le. Jgr Z 23 nA.
Distance piece.
Cork washer
Gaine Zdlg 36.
Flash channel.
Exploder container.
Amatol 40/60 cast.

Fig. 20

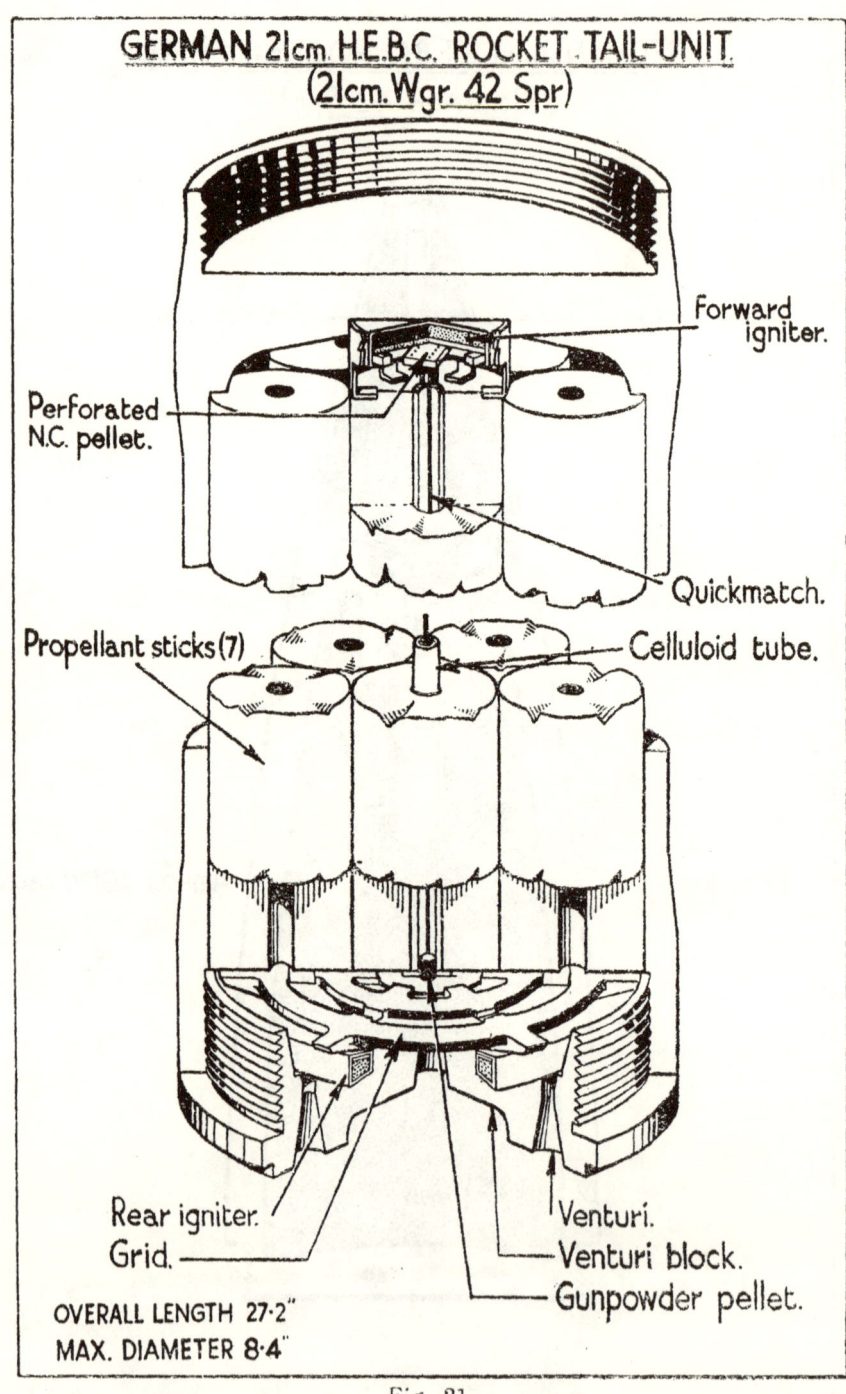

Fig. 21

The complete round consists of the following components :—
  Shell H.E.B.C. filled amatol 40/60 ;
  Nose fuze le Jgr. Z. 23 nA. Hbgr. Z. 35 K., or Zt. Z. S/30 ;
  Gaine Zdlg. 36 Np. ;
  Tail tube with propellant charge and ignition system ;
  Venturi block with grid ;
  Electric ignition fuze.

Shell (Fig. 20)

The general shape of the shell is a truncated ogive of high crh. from the base. The filled shell, including the fuze weighs approximately 98 lb. The bursting charge, indicated by the numeral " 13 " stencilled on the body, is amatol 40/60 and weighs approximately 22·4 lb. The shell body is of steel with a solid base ; it is screwthreaded externally at the base for the attachment of the tail tube and, on the shoulder to receive the ballistic cap. The overall length of the shell including the threaded portions, is 15·8 inches. The diameters at the base and shoulder are 8·38 inches and 5·3 inches. The shoulder is screwthreaded internally to receive an adapter, which in turn takes the fuze and a steel exploder container. A steel distance piece 1·5 inches long and 1·54 inches in diameter, with a central flash hole 0·32 inch in diameter, separates the fuze from the gaine. A cork washer attached to the end of the distance piece is in contact with the fuze.

The ballistic cap, approximately 8·5 inches long, is of thin metal and shaped to the general contour of the shell body. The nose is flat and bored centrally to receive a screwed adapter supporting the upper end of a wooden rod. The wooden rod rests on the top of the fuze and forms an extension to the striker. It is retained by spinning the nose of the adapter over a flat retaining disc. The base of the cap is screwthreaded internally for attachment to the shell body.

Fuze and Gaine

The fuzes le Igr. Z. 23 nA and Hbgr. Z. 35 K., and the gaine Zdlg. 36 Np. are described as separate items in this pamphlet. Fuse Zt. Z. S/30 is described in Pamphlet No. 8, page 11.

Tail Unit (Fig. 21)

The tail unit has an overall length of 27·2 inches and weighs approximately 144¼ lb. filled. The main parts comprise a steel tail tube, steel venturi block, propellant charge, steel grid and an ignition system.

The tail tube weighing 87 lb. 2 oz. is cylindrical in shape, 26·6 inches long and 8·27 inches in diameter, except at the venturi end and near the head end, where the diameters are slightly increased. The wall is machined inside and out to a thickness of 0·394 inch.

The head end is closed by a diaphragm formed in the tube. A cavity approximately 1·25 inches deep in front of the diaphragm is screwthreaded internally to receive the base end of the projectile. The opposite face of the diaphragm is recessed to the shape of a saucer approximately 7 inches in diameter across the rim. The centre of this recess has a cavity partially to accommodate the forward igniter. Internally, the venturi end is screwthreaded to receive the venturi block.

The venturi block is solid, flanged and screwthreaded externally for insertion in the end of the tube. Twenty-two venturis, equally spaced on a circle 6·08 inches in diameter, have their axis inclined at an angle of 16 degrees to the axis of the rocket, so that the escaping gases cause the round to rotate in flight. An axial venturi in the centre of the block extends into a boss formed on the front face. The boss is screwthreaded externally to take a grid supporting the base end of the propellant charge. The venturis have a throat diameter of 0·362 inch and open at an inclusive angle of 26 degrees.

The propellant charge, weighing approximately 39 lb. 9¼ oz., consists of six sticks of tubular propellant surrounding a seventh. Each is 21·66 inches long, 2·45 inches and 0·32 inch external and internal diameters respectively. The forward end of the outer sticks is held by a shoulder formed in the tube near the rim of the saucer-shaped recess; the centre stick is held by the lugs of the igniter holder. A preliminary chemical analysis shows that charge is basically approximately 59 per cent. nitrocellulose and 37 per cent. diglycoldinitrate.

The grid supporting the charge at the venturi end is shaped to form three concentric circles on a web of six radial arms; it weighs 1 lb. 10 oz., and its overall diameter is 7·3 inches. The centre circle is screwthreaded internally to enable the grid to be screwed to the boss on the front face of the venturi block. The arms are chamfered on the base side towards their outer ends.

The ignition system comprises a forward igniter, a length of primed celluloid tubing and a rear igniter.

The forward igniter, which weighs 3 ozs., is housed within a steel split ring, forming an igniter holder with six lugs bent inwards to form distance pieces. The igniter consists of a pellet of pressed grey powder composed of aluminium and barium nitrate under a thin black layer of compressed zirconium metal and potassium nitrate. Across the face of this thin layer is a perforated strip of nitrocellulose. The filling is contained in a flat circular aluminium container 2·57 inches in diameter, and retained by turning over the lip of the container on to a ring of compressed paper.

The rear igniter, accommodated immediately in front of the venturis and in rear of the grid, is in the form of a celluloid " U " section ring filled gunpowder. The external and internal diameters of the ring are 6·78 inches and 5·6 inches respectively, and its outer and inner thickness ·5 and ·33 inches respectively.

The celluloid tubing is 22·35 inches long and is accommodated in the central stick of propellant. It contains quickmatch and at each end a perforated gunpowder pellet. The forward pellet is accommodated centrally within the lugs of the igniter holder. The rear pellet protrudes slightly from the base of the central stick.

The outer venturis are closed by a thin ring of aluminium foil, 6·76 inches in diameter, fitted against the front face of the venturi block. The centre venturi is closed by a thin disc of aluminium foil 1·96 inches in diameter fitting in the cavity in the front face of the grid.

Fuze

This is an electric ignition fuze and is described elsewhere in this pamphlet. It is pushed into one of the venturis when the rocket is loaded into the projector for firing.

Action
  Tail Unit

An electric current fires the ignition fuze and the flash ignites the rear igniter, which in turn ignites the length of quickmatch in the celluloid tube and the forward igniter. The latter ensures complete ignition of the propellant charge. Pressure set up inside the tube and escaping through the venturis propels the rocket forward. The inclined venturis cause the rocket to rotate and stabilize it in flight.

  Head Unit

On impact, the ballistic cap is shattered and the wooden rod is driven in to fire the fuze. The flash from the fuze passes through a flash channel in the steel distance piece to detonate the gaine which in turn detonates the shell filling.

GERMAN 30 cm. H.E. ROCKET
30 cm. Wurfkörper. Spreng
(Figs. 22 and 23)

This is a self-propelled base venting rocket, stabilized in flight by rotation caused by the inclined venturis. The rocket is fired electrically from the 30 cm. Nebelwerfer 42. The complete round weighs approximately 277 lb. Its overall length is 46·5 inches; the maximum diameter of the bomb is 11·75 inches and of the tail tube 8·56 inches. The centre of gravity is 21·75 inches from the base. The exterior of the rocket is painted deep olive green and the stencilling in black, except the abbreviated nomenclature of the round " 30 cm. WK Spr." near the nose end, which is stencilled in white. External markings on the round are shown in Fig. 22.

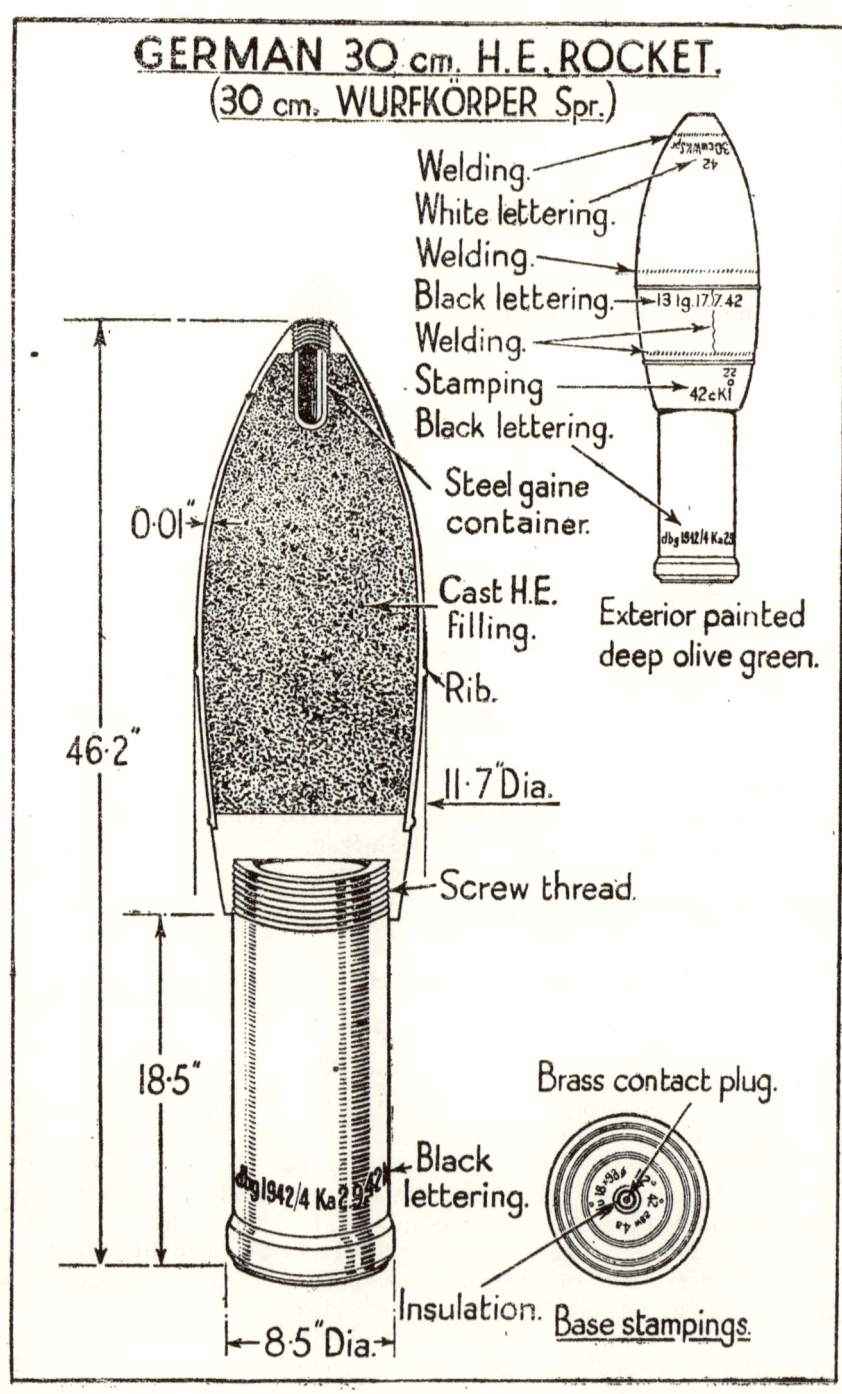

Fig. 22

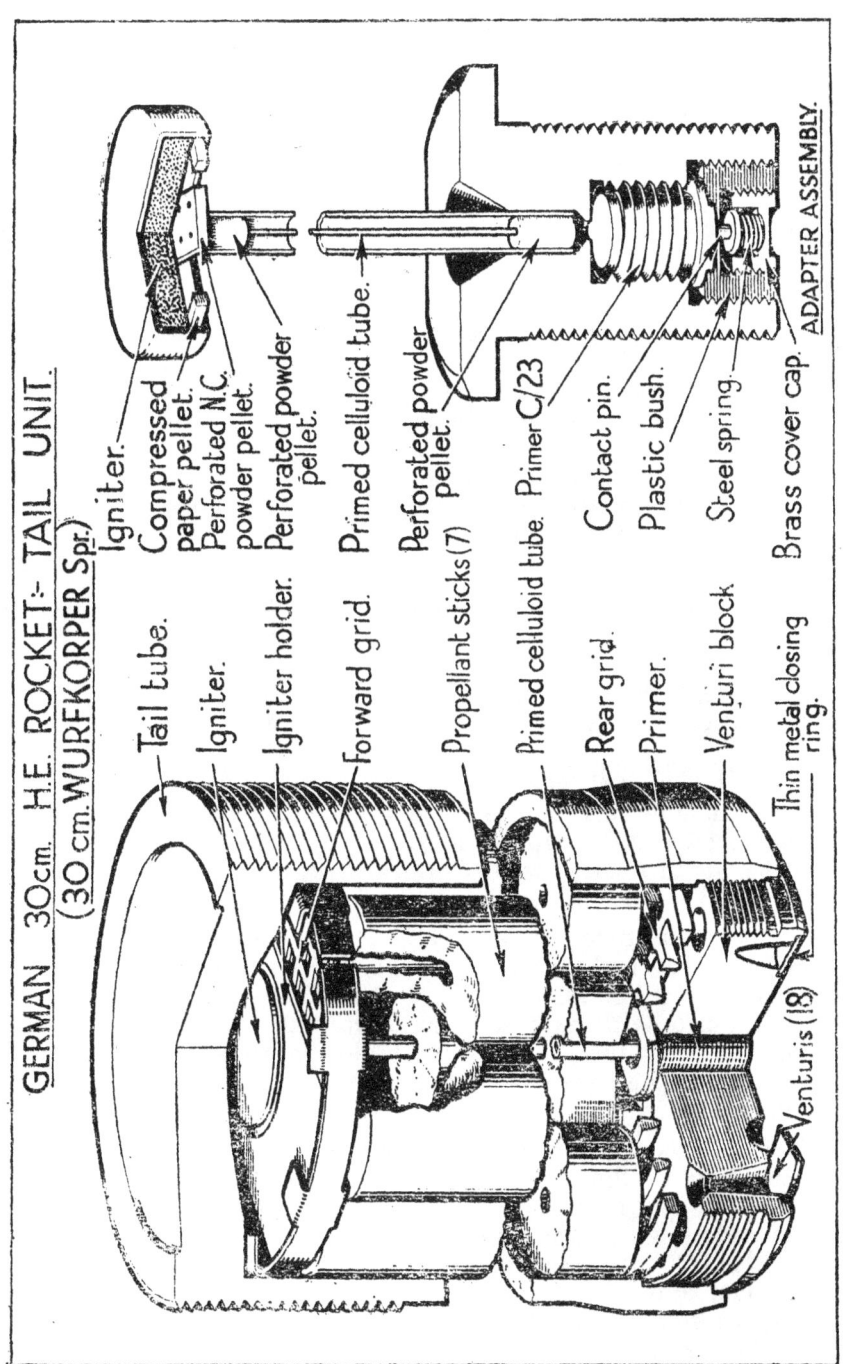

Fig. 23

The complete round consists of the following components :—
  Shell H.E. filled amatol 40/60 ;
  Nose fuze ;
  Gaine ;
  Tail unit with propellant charge and ignition system ;
  Primer electric C/23 with contact unit.

Shell (Fig. 22)

The weight of the filled shell without fuze and gaine is approximately 147 lb. The general shape of the shell is an ogival head of medium crh., whilst the remainder of the body is streamlined towards the base. The body consists of two wall sections and a solid nose and base welded together circumferentially. The wall sections are approximately 0·1 inch thick and welded at their maximum diameter. A rib is formed in the body behind the weld. The rear section is also welded longitudinally. The nose section is screwthreaded internally to receive a nose fuze and a steel exploder container. In transport the nose is closed by a plastic plug. The base is screwthreaded internally for the attachment of the tail tube.

The bursting charge as indicated by the numeral " 13 " stencilled on the body, is amatol 40/60.

Fuze and Gaine

Details of the fuze and gaine are not yet available.

Tail Unit (Fig. 23)

The tail unit has an overall length of 22·6 inches and weighs approximately 129 lb. 10 oz. filled. It consists essentially of the tail tube, venturi block, propellant charge, two grids, and an ignition system with electric primer and spring contact.

The steel tail tube is cylindrical in shape, 8·34 inches in diameter except near the base, where a humped band is formed to increase the diameter to 8·56 inches. The wall of the tube is 0·43 inch thick and machined both externally and internally. The head end is closed and screwthreaded externally for insertion into the shell. Internally, at the base end, it is screwthreaded and provided with a shoulder to receive the venturi block.

The solid venturi block is flanged and screwthreaded externally for insertion in the venturi end of the tube. Eighteen venturi holes are drilled in the block with their centres on a circle of 3·05 inches radius. The throat diameter of each venturi is approximately 0·365 inch and the exit diameter approximately 0·82 inch. The axis of the venturis are inclined at an angle of 12 degrees 42 seconds, so that the effluent gases cause the round to rotate in flight. A thin tin plate ring fits into a circular recess in the base of the block and seals the venturis exits. The base of the block is stamped thus: "Wu 18 × 9·3 ϕ 12° 42asw 5a" indicating the number, the throat diameter in millimeters, and the inclination angle of the venturis. The centre of the venturi block is bored and screwthreaded to receive an adapter. To facilitate assembly in the tail tube, two key recesses are provided in the base of the block.

The propellant charge weighs 33 lb. and consists of six tubular sticks of propellant surrounding a seventh. Each is 18·4 inches long, 2·45 inches and 0·32 inch external and internal diameters respectively. Preliminary chemical analysis has shown that the charge contains 59·9 per cent. nitrocellulose and 35·4 per cent. diglycoldinitrate.

The cast steel grid supporting the venturi end of the charge is shaped to form four concentric circles on a web of six arms. The webs are chamfered on their base side towards their outer ends. The grid is bolted to the venturi block by the primer adapter.

The grid at the forward end of the charge consists of a split ring with eight pairs of lugs and a sheet of metal mesh. Each pair of lugs is bent inwardly to a "U" shape in section, and the metal mesh is welded to two adjacent lugs diametrically opposite the split. In addition to spacing, the grid prevents the igniter being crushed by the charge.

The ignition system consists of an igniter, a primed celluloid tube and an electric primer unit.

The igniter, which weighs 3 oz., is housed within the split ring and held centrally opposite the central stick of propellant by an igniter holder. The igniter consists of a pellet of pressed black powder with a perforated N.C. powder pellet in strip form across its surface. The pellets are contained in a flat aluminium capsule and retained by turning over the lip of the capsule on to a compressed paper ring. The igniter holder is a plate of sheet steel, 0·04 inch thick, 3·95 inches wide and 7·42 inches long, radiused at the ends to suit the curvature of the grid split ring. A hole 2·6 inches in diameter is bored in the centre of the holder, and a ⌐ ring-shaped in section, to hold the igniter, fits into the hole and is spot welded at six points.

The celluloid tube, accommodated in the central stick of propellant, is 19·7 inches long and 0·23 inch and 0·0394 inch external and internal diameters. It contains quickmatch and at each end a perforated gunpowder pellet. The forward pellet is accommodated centrally within the forward grid and is in contact with the igniter. The pellet at the other end is accommodated in the vent of the primer adapter.

The primer adapter, which also bolts the rear grid to the venturi block, is screwthreaded externally and flanged at one end; it is inserted into the venturi block from the front, and two flats formed on the periphery of the flange enable it to be screwed home with the aid of a spanner. The adapter is bored centrally in four dimensions with a coned vent at the flanged end. Two chambers formed in the base end of the boring are screwthreaded to receive the electric primer and contact unit respectively, and the others form a necked flash channel which houses the gunpowder pellet and the rear end of the inflammable tube. The smaller of the threaded borings, formed in the centre of the adapter, received the electric primer " C/23 " described as a separate item in this pamphlet. The larger boring nearer the base houses the spring contact unit. The unit consists of a plastic insulating bush, steel pin, steel spring and brass cover cap. The bush is cup-shaped with a central hole bored in its base; it is screwthreaded externally for insertion in the adapter and internally to receive the cover cap. The contact pin is a stem with circular flat head and is inserted inside the bush so that the stem portion protrudes through the hole to make contact with the contact plug of the electric primer. The pin is kept in contact by the spring held in compression between the head of the pin and cover cap. The cover cap is cup-shaped to house the spring and head of the pin, and screwthreaded externally for insertion in the bush.

Action of Tail Unit

An electric current passes through the contact unit and fires the primer. The flash from the primer ignites the gunpowder pellet and quickmatch in the celluloid tube, which in turn ignites the igniter, bringing about ignition of the propellant. Pressure set up in the tail blows off the thin metal ring closing the venturi exits, propels the projectile, which, owing to the inclination of the venturis, is rotated.

GERMAN 4 cm. CARTRIDGE Q.F. H.E.
(BOFORS TYPE)
(4 cm. Sprgr. Patr. 28 Flak)

The cartridge is used in the 4 cm. Flak 28 A.A. gun of the Bofors type. The length of the complete round is 17·7 inches and its weight is approximately 4 lb. 4 oz. The shell is painted yellow and stencilled in red, and has a red band painted above the driving band. The stencilling on the shell body, above the red band, includes the numeral " 1 " indicating the nature of filling. The shell is fitted with a single copper driving band.

The fixed Q.F. cartridge consists of the following components :—

Shell H.E. filled T.N.T. and tracer composition ;
Self-destroying detonator ;
Tracer igniter ;
Fuze Kz 38 with gaine
Brass case or steel case coated with brass ;
Propellant charge of double base propellant with a nitrocellulose igniter ;
Primer percussion C/33 St.

The complete round may be fired from the British 40 mm. equipments.

Shell (Fig. 24)

The shell is substantially the original 1938 Bofors shell, modified to give a greater certainty of detonation and a longer range before self-destruction. Externally, it differs in that it tapers more sharply in its streamlined base. The weight of the shell, filled and fuzed, is 2 lb. 1 oz. 1 dr., and its overall length when fuzed is 7 inches. A cannelure is formed in rear of the driving band to enable the shell to be attached to the cartridge case. The body, at about its shoulder, is screwthreaded internally to receive an adapter, which tapers to form the forward end of the shell body and to receive the gaine and fuze.

Internally the shell has two compartments containing the main H.E. filling and the tracer composition respectively, and a recess in the base containing the tracer igniter, all of which are interconnected. The bursting charge, in the forward compartment forming the front half of the shell body, consists of a 1 oz. 5½ dr. pellet of pressed biscuit T.N.T. contained in a varnished carton with a cavity in the base to receive the self-destroying detonator and another in the forward end to receive the gaine. The cavity in the base is fitted with a brass liner.

At the base of the bursting charge is a steel plate with a central flash hole, which fits against an internal shoulder formed by the reduced diameter of the shell cavity and separates the bursting charge from the tracer composition. Between the steel plate and the bursting charge is a box cloth washer.

The self-destroying detonator consists of 8·5 grains of CE. over 6 grains of lead azide/lead styphnate (63/37) contained in an aluminium cylinder with a flash hole closed by a fabric disc in the base. The forward end of the cylinder is closed by an aluminium washer retained in position by turning over the mouth of the cylinder.

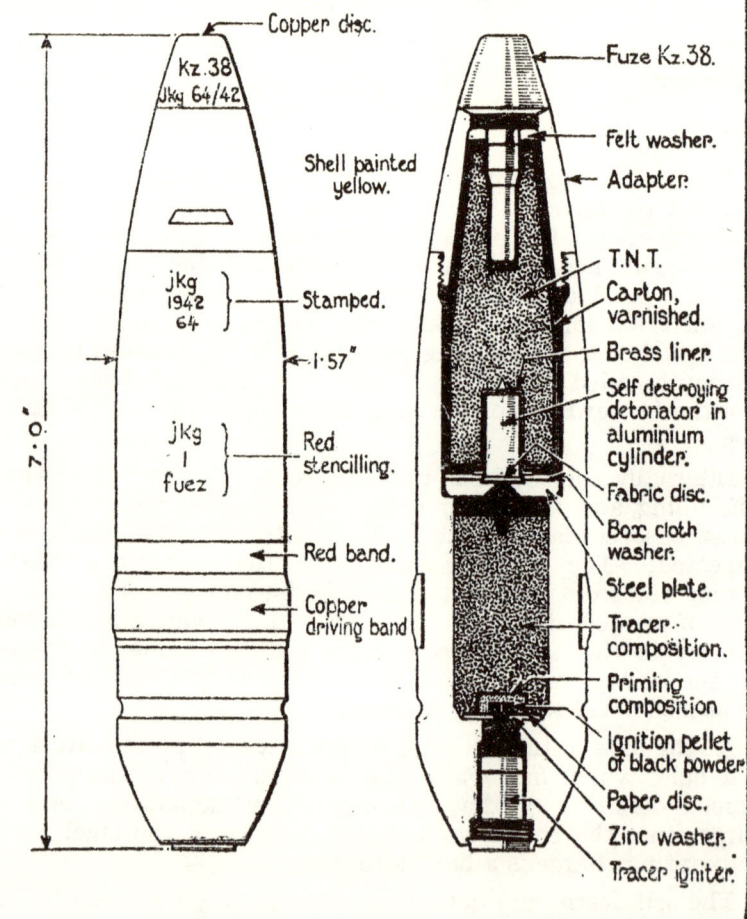

Fig. 24

The tracer composition, pressed directly into a compartment of slightly less diameter and length than that containing the bursting charge, is a purplish composition weighing 1 oz. 2¼ dr. and consisting of resin (containing shellac) 10·6 per cent., magnesium-aluminium alloy (90/10) 30 per cent., strontium nitrate 59 per cent., volatile matter 0·4 per cent. The inclusion of aluminium increases the brilliance of the trace and the time of burning. The tracing composition is primed with a pellet of priming composition similar in appearance to the main tracer filling but containing small grains of black powder. The priming composition surrounds a smaller perforated ignition pellet of fire-grained gunpowder.

The analysis of the pellets is as follows :—

	Priming pellet.	Ignition pellet.
Sulphur	9·0 per cent.	4·4 per cent.
Charcoal	12·9	6·6
Potassium nitrate	71·6	44·5
Magnesium	1·5	15·8
Barium peroxide	3·0	25·7
Resin	1·0	2·0
Volatile matter	1·0	1·0

The pellets are contained in a cylindrical recess formed in the base of the tracing composition. The compartment is closed at its base end by a zinc washer and paper disc varnished to its upper surface.

The trace burns for approximately 9·4 seconds and averages 2,680 candle power when spinning at 30,000 r.p.m.

Tracer Igniter (Fig. 25)

The tracer igniter is of the same design as that in the British 40 mm. shell Mark I. The body has a flange at its base and is recessed and screwthreaded to receive a pellet holder which screws on to a copper sealing washer. The recess is prepared at the base to accommodate a percussion cap kept in position by a brass washer and, at its forward end, a hammer held off the cap by a stirrup spring. The pellet holder is recessed at the top and bottom to form a chamber for the magazine and upper part of the hammer respectively, the recesses are connected by a central channel. The magazine contains a gunpowder pellet and is closed at the top by an aluminium washer with a paper disc on its under side. The central channel is filled with pressed powder and has a rough surface to prevent the movement of the filling. The hammer is cylindrical with its lower part shaped to form a round headed striker and is prepared with two flash channels leading to a recess at its upper end. The stirrup spring is cylindrical with flanges formed by cutting the metal and bending it inwards to support the hammer and keep it clear of the cap until the gun is fired. The cap contains detonator composition.

GERMAN TRACER IGNITER FOR 4 cm. SHELL.

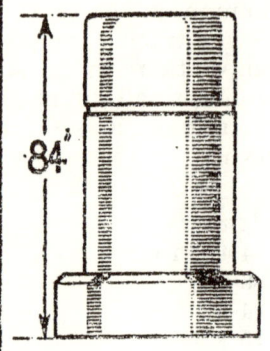

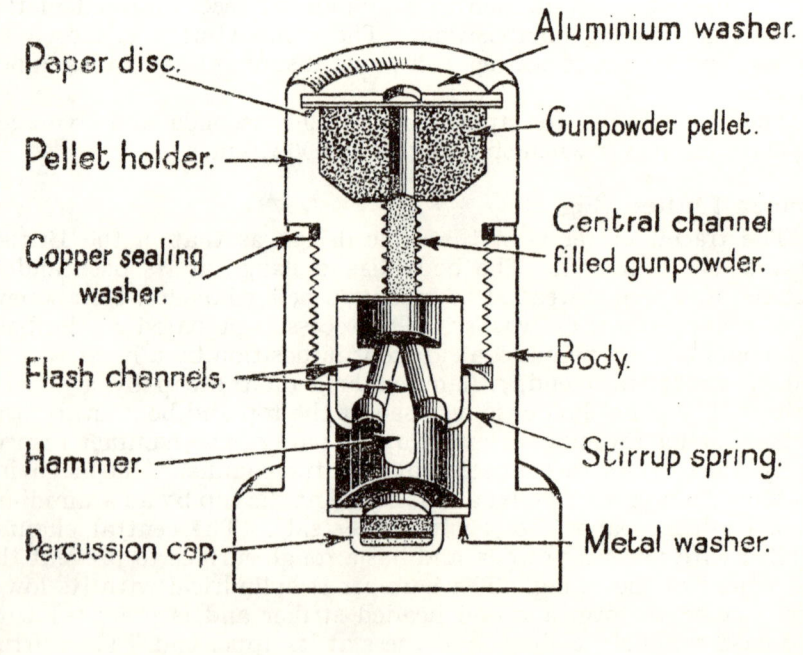

Fig. 25

Action

On acceleration, the hammer sets back, overriding the flanges of the stirrup spring and strikes the percussion cap. The flash from the cap passes through the flash channels in the hammer and fires the gunpowder in the central channel and magazine which in turn ignite the tracer composition *via* its ignition and priming pellets. The trace enables the flight of the shell to be observed. If the fuze has not already functioned the trace eventually detonates the self-destroying detonator, which in turn detonates the bursting charge of the shell.

Fuze Kz 38 with Gaine

This fuze and gaine are described elsewhere in this pamphlet under the above heading.

Cartridge Case

The cartridge case is 12·25 inches in length and is of the normal Bofors type with a groove immediately in front of the flange and another in the base. Cases may be identified by the stamping " 4 cm. 28 " or " 4 cm. 28 St." on the base. The former is a brass case and the latter a steel one with a coating of brass. The case is secured to the shell by indenting it into a cannelure near the base of the shell.

Propellant Charge

The propellant charge indicated by the black stencilling on the case is 275 g (9 oz. 11 dr.) Ngl. RP — (234 · 2, 2/0, 8), and is fitted with an igniter. The analysis of the propellant and igniter of the round examined are as follows :—

Propellant—Nitrocellulose 55·92 per cent., nitroglycerine 34·07 per cent., diphenylamine 0·76 per cent., ethyl-centralite 5·15 per cent., unidentified oily substance 4·10 per cent.

Igniter—Nitrocellulose 93·25 per cent., nitroglycerine 3·31 per cent., diphenylamine 1·27 per cent., graphite 0·52 per cent., volatile matter 1·65 per cent.

Primer

The percussion primer of the round examined was the C/33 St. The primer is of the same design as the C/33 described in Pamphlet No. 7, page 45, except that the body is of steel instead of brass.

Variations in Shell Fillings and Primers

In addition to the round described, two other rounds have been examined and found to be fitted with shell containing a bursting charge of 1 oz. 5¾ dr. of PETN/Wax (90/10) which will give an increased damage effect. The shell did not bear a nature of filling number but the red stencilling on the shell body above the driving band included the abbreviation " NP " denoting the nature of filling.

The tracer composition of these rounds did not contain aluminium; its analysis is as follows :—

> Strontium nitrate 50·8 per cent., magnesium metal 31·2 per cent., barium nitrate 3·2 per cent., resinous matter 14·8 per cent.

One cartridge case was of the original Bofors design and was fitted with the larger German percussion primer C/12 described in Pamphlet No. 4, page 10, and the other with the smaller German percussion primer C/13 described in Pamphlet No. 7, page 44.

GERMAN 8·8 cm. Flak 41 CARTRIDGE Q.F. A.P.C.B.C./T.
(8·8 cm. Pzgr. Patr. 39 Flak 41)

(Fig. 26)

This cartridge is used in the 8·8 cm. Flak 41 gun presumably in the anti-tank role. The weight of the complete round is 46 lb. 6 oz. and its overall length is 45.8 inches.

The fixed Q.F. cartridge consists of the following components :—

> Armour piercing shell fitted with penetrative and ballistic caps and filled H.E.
> Fuze Bd. Z. 5127 with tracer ;
> Brass case ;
> Propellant charge of double base flashless propellant ;
> Primer electric C/22.

Shell

The weight of the shell, filled and fuzed, is approximately 22 lb., 5¼ oz. The exterior is painted black with white tip on the ballistic cap and stencilled in red. It is fitted with two driving bands of soft iron. Two cannelures for the attachment of the case are formed behind the driving band and the base is screwthreaded to receive the fuze with tracer. A penetrative cap is attached to the body and a ballistic cap fits over the penetrative cap. The cavity for the bursting charge is smaller than that in the shell for the Flak 18 and 36 described in Pamphlet No. 6, page 22. The bursting charge, identified by the numerals "92" stencilled on the shell body is cyclonite/wax (90/10) and is approximately 2 oz. 2 dr. in weight.

The weight of the empty projectile is 21 lb. 7¼ oz.

Fuze Bd. Z. 5127 with Tracer

Details of this fuze are given in Pamphlet No. 12 which includes a drawing of the fuze and tracer.

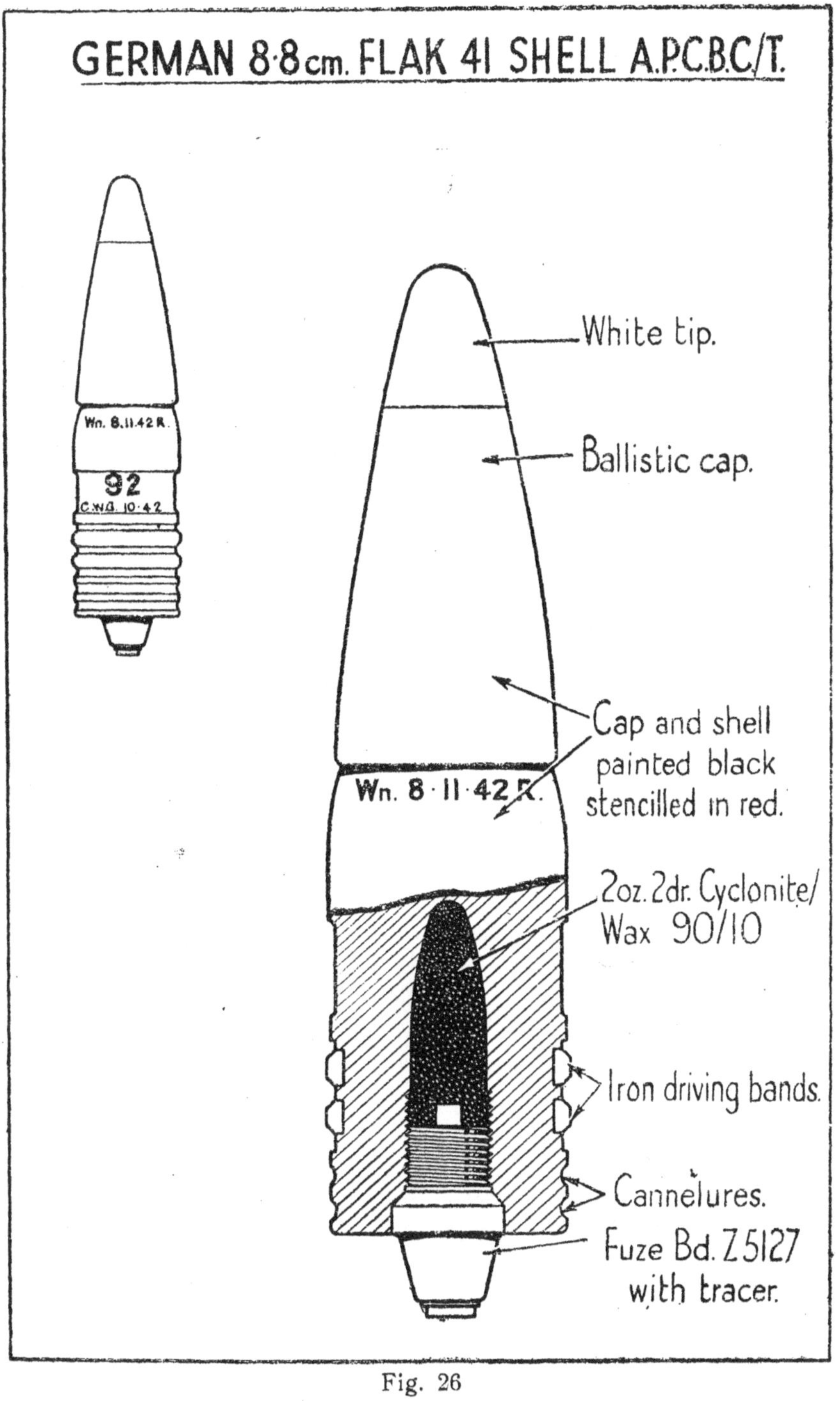

GERMAN 8·8 cm. FLAK 41 SHELL A.P.C.B.C/T.

- White tip.
- Ballistic cap.
- Cap and shell painted black stencilled in red.
- 2 oz. 2 dr. Cyclonite/Wax 90/10
- Iron driving bands.
- Cannelures.
- Fuze Bd. Z 5127 with tracer.

Wn. 8.11.42 R.
92
C.W.G. 10·42

Wn. 8·11·42 R.

Fig. 26

Cartridge Case

The cartridge case is of brass, 33·8 inches in length, with a slight increase in taper near its mouth. The case is stamped in the base " 8·8 cm. Flak 41 " and, unlike most German cases, does not appear to have been allotted a case model number.

Propellant Charge and Igniter

The flashless propellant charge, as indicated by the stencilling on the case, weighs 11 lb. 15 oz., and is made up of tubular cords consisting basically of nitrocellulose diethylene glycoldinitrate with the addition of nitroguanidine. The bag containing the charge has an igniter containing 20 grams of nitrocellulose powder sewn to its base. The weight, nature and size of the propellant charge is stencilled on the case as follows :—" 5·415 Kg. Gu. R.P.–7, 5– (740 – 4,7/1, 5)." The case is also marked " Tropen," indicating the charge to be suitable for hot climates.

Primer Electric C/22.

Details of this primer are given in Pamphlet No. 4, page 31, and Fig. 20.

GERMAN 12·8 cm. Flak 40 CARTRIDGE Q.F.
H.E. Fuzed Zt Z.S/30
(12·8 cm. Sprgr Patr. L/4·5)

(Fig. 27)

This fixed Q.F. cartridge is used in the Flak 40 A.A. gun, which may be static, mobile or mounted on railed vehicles. The length of the complete round is 58·5 inches and it weighs approximately 100 lb. 4 oz.

The complete round consists of :—
- Shell H.E. filled amatol 40/60 ;
- Fuze Zt Z.S./30 ;
- Brass cartridge case ;
- Propellant case of tubular Digl ;
- Primer electric C/22.

Shell

The shell is painted yellow, stencilled in black and fitted with two iron driving bands. The bursting charge of amatol 40/60 cast, indicated by the numeral " 13 " stencilled on the shoulder, weighs 7 lb. 7½ oz. An exploder container of the larger size, for the 36 gaine, is screwed into the fuze hole. The weight of the shell, filled and fuzed, is 57 lb. 5 oz.

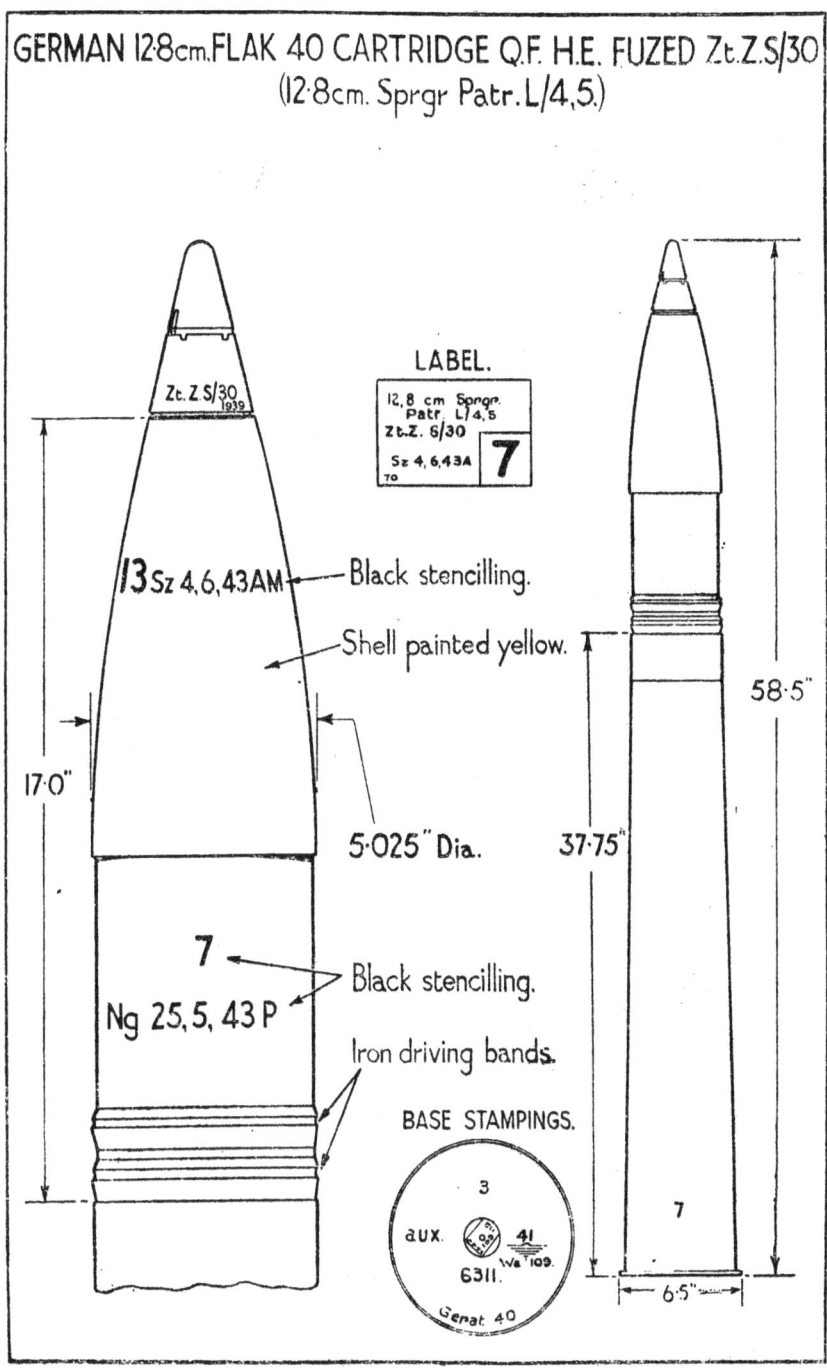

Fig. 27

Gaine

The gaine used below the time fuze is the Zdlg. 36 Np described as a separate item in this pamphlet.

Fuze

The fuze, Zt. Z.S/30, is described in Pamphlet No. 6 page 11, and Pamphlet No. 8, page 11.

Cartridge Case

The brass cartridge case is the normal flanged type, 37·75 inches long with a slight increase in taper towards the mouth. The case model number " 6311 " and " Gerat 40 " is stamped in the base. The effective capacity of the case is 864 cubic inches.

Propellant Charge

The propellant charge consists of 9·62 Kg (21 lb. 3½ oz.) Digl RP – KN – (850 · 5,6/2) fitted with an igniter consisting of 239 grams of nitrocellulose composition.

Primer

Details of the electric primer C/22 are given in Pamphlet No. 4, page 31.

www.ingramcontent.com/pod-product-compliance
Lightning Source LLC
Chambersburg PA
CBHW032012080426
42735CB00007B/578